ISO 14000
Guide

Environimental Engineering Books

ALDRICH • *Pollution Prevention Economics: Financial Impacts on Business and Industry*

AMERICAN WATER WORKS ASSOCIATION • *Water Quality and Treatment*

BAKER, HERSON • *Bioremediation*

BRUNNER • *Hazardous Waste Incineration*, Second Edition

CALLAHAN, GREEN • *Hazardous Solvent Source Reduction*

CASCIO, WOODSIDE, MITCHELL • *ISO 14000 Guide: The New International Environmental Management Standards*

CHOPEY • *Environmental Engineering in the Process Plant*

COOKSON • *Bioremediation Engineering: Design and Application*

CORBITT • *Standard Handbook of Environmental Engineering*

CURRAN • *Environmental Life-Cycle Assessment*

FIKSEL • *Design for Environment: Creating Eco-efficient Products and Processes*

FREEMAN • *Hazardous Waste Minimization*

FREEMAN • *Standard Handbook of Hazardous Waste Treatment and Disposal*

FREEMAN • *Industrial Pollution Prevention Handbook*

HARRIS, HARVEY • *Hazardous Chemicals and the Right to Know*

HARRISON • *Environmental, Health, and Safety Auditing Handbook*, Second Edition

HAYS, GOBBELL, GANICK • *Indoor Air Quality: Solutions and Strategies*

JAIN, URBAN, STACEY, BALBACH • *Environmental Impact Assessment*

KOLLURU • *Environmental Strategies Handbook*

KOLLURU • *Risk Assessment and Management Handbook for Environmental, Health, and Safety Professionals*

KREITH • *Handbook of Solid Waste Management*

LEVIN, GEALT • *Biotreatment of Industrial and Hazardous Waste*

LUND • *The McGraw-Hill Recycling Handbook*

MAJUMDAR • *Regulatory Requirements of Hazardous Materials*

ROSSITER • *Waste Minimization Through Process Design*

SELDNER, COETHRAL • *Environmental Decision Making for Engineering and Business Managers*

SMALLWOOD • *Solvent Recovery Handbook*

WILLIG • *Environmental TQM*

ISO 14000 Guide

The New International
Environmental Management
Standards

Joseph Cascio

Gayle Woodside

Philip Mitchell

McGraw-Hill

New York   San Francisco   Washington, D.C.   Auckland   Bogotá
Caracas   Lisbon   London   Madrid   Mexico City   Milan
Montreal   New Delhi   San Juan   Singapore
Sydney   Tokyo   Toronto

Library of Congress Cataloging-in-Publication Data

Cascio, Joseph.
    ISO 14000 guide : the new international environmental management
standards / Joseph Cascio, Gayle Woodside, Philip Mitchell.
        p.   cm.
    Includes bibliographical references.
    ISBN 0-07-011625-3
    1. ISO 14000 Series Standards.   I. Woodside, Gayle.
II. Mitchell, Philip.   III. Title.
    TS155.7.C37   1996
    658.4′08—dc20   96-4168
                                                          CIP

McGraw-Hill

A Division of The McGraw·Hill Companies

    3 4 5 6 7 8 9 0   DOC/DOC   9 0 1 0 9 8 7

ISBN 0-07-011625-3

*The sponsoring editor for this book was Zoe G. Foundotos, the editing super-
visor was Patricia V. Amoroso, and the production supervisor was Donald F.
Schmidt. It was set in Palatino by Donald Feldman of McGraw-Hill's
Professional Book Group composition unit.*

Printed and bound by R. R. Donnelley & Sons Company.

Contents

Preface

ISO 14000 embodies a new approach to environmental protection. In contrast to the prevailing command-and-control model, it challenges each organization to take stock of its environmental aspects, establish its own objectives and targets, commit itself to effective and reliable processes and continual improvement, and bring all employees and managers into a system of shared and enlightened awareness and personal responsibility for the environmental performance of the organization. This new paradigm relies on positive motivation and the desire to do the right thing, rather than on punishment of errors. Over the long term, it promises to establish a solid base for reliable, consistent management of environmental obligations.

Recent industrial accidents, some entailing significant human and environmental harm, have proved that regulatory compliance is not enough to ensure against environmental degradation. As it became clear that compliance was not a complete prescription for environmental protection, an awareness arose that a more proactive system was needed. ISO 14001, the foundation of the entire ISO 14000 series, is such a proactive environmental protection strategy in which regulatory compliance is but one of the elements of a more inclusive and all-encompassing approach.

ISO 14001, the environmental management system (EMS) standard, provides a framework to direct the use of organizational resources to the full breadth of actual and potential environmental impacts through reliable management processes and a base of educated and committed employees. Regulatory compliance is now a normal result of this management strategy, along with awareness, sensitivity, and preparedness, greater reliability and consistency in meeting environmental objectives, and greater confidence in the organization's ability to prevent accidents.

After decades of focusing on compliance with government regulations, however, the regulated and regulating communities will need to engage in some rethinking to look beyond compliance as the measure of an organization's environmental achievement. Compliance will, of course, lose none of its importance in an organization's operations. But it would be shortsighted to view ISO 14001 as merely a tool to achieve compliance, and those who insist on doing so will incur the costs of implementing the EMS without reaping its full benefits. It is imperative, therefore, that everyone involved with ISO 14001 understand its wider purpose and avoid trivializing it by setting its value only with reference to its impact on regulatory compliance. ISO 14001 is a significant and consequential development in our ability to protect and preserve the environmental resources of our planet—transcending the regulatory compliance approach—and must be valued accordingly by both users and regulators.

There should be no illusion that ISO 14001 will be easy to implement. Even organizations with sophisticated environmental programs will find ISO 14001 challenging. The organization must inventory and then assess all environmental aspects of its operations, products, and services. Regulations may apply to many of these, but they are not likely to apply to all. The standard calls for a system that produces reliable and effective management. While regulations call for compliance, they generally do not include requirements for management systems. ISO 14001 expects all employees to be trained and competent in handling the environmental consequences of their work. This requires the infusion of environmental awareness and attitudes in all workers. Broadly, the result over time is a shift in culture to one that is as sensitive to the environment as to production schedules and product design. Few regulations require such far-reaching changes in the mental attitudes of all employees.

It is also true, however, that the diffusion of environmental responsibility from the environmental engineering function to all employees in the enterprise will be the biggest challenge and one that, in the short term, may carry some risk of administrative noncompliance, as employees learn documentation and other recordkeeping tasks. But since the goal is to broaden the organizational base of environmental responsibility, we must be willing to accept the possibility of these types of errors during the early phases of implementation. Thus, it needs to be understood that conformance to ISO 14001 is not likely to result in an immediate change in the organization's compliance posture.

This book was written for organizations that set their sights on conformance to ISO 14001. It is designed to provide insight on the ISO 14000 concept and guidance for its implementation. Part 1 is the overview, explaining the conceptual basis of the ISO 14000 series and describing how and why it was developed. Part 2 provides valuable, practical information about implementing the key elements of an EMS. My coauthors, Gayle Woodside and Phil Mitchell, address each key element of the ISO 14001 standard, in the same order that the standard does, so that the organizational manager can readily understand its requirements. Numerous tables and figures throughout the book clarify important points addressed in the standard. The three of us understand the great journey an organization undertakes with ISO 14001. Gayle and Phil join with me in wishing all of you well who undertake that journey.

Joe Cascio

Acknowledgments

My thanks extend to Diana J. Bendz, Director of Integrated Safety Technology at IBM, for her continuous support of my role in the ISO process; to Lawrence L. Wills, IBM Director of Standards and ANSI Chairman, for his constant encouragement; and to Mary Anne D. Lawler, Director of Standards Relations at IBM, for her much appreciated guidance and counsel over the last four years. Additional thanks to all the participants in the USTAG, the members of CAG, and all colleagues in TC 207 who have contributed to my understanding of management systems and tools. Finally, thanks to my family—Keith, Ted, Tom, and Pat—for their enthusiastic cheerleading and wonderful patience with a father and husband who was mostly missing during the summer and fall of 1995.

Joe Cascio

A note of thanks to the following people for their technical help and continued support: John Woodside, John Prusak, Jeanne Yturri, Dianna Kocurek, and IBM-Austin's Environmental Engineering Department—Kelvin Langlois, Bonnie Blam, David Dalke, Chris Bauer, Bob Tassan, Jeff Erb, Barbara Salomon Casey, Stuart Hurwitz, Ken Takvam, Rich Reich, and Chau Vo. And, of course, thanks to my husband and friend, Bruce Almy, for all his encouragement and support.

Gayle Woodside

My thanks to Jim Dumanowski, Doug Williams, June Andersen, and Elizabeth Zimmermann for their insights and comments. Thanks also to my family—Polly, Tracy, and Scott—for their patience and support during the writing and editing of this book. A final note of gratitude to some of those who have shaped my thinking in the environmental management area: Don Fast, H. Ray Kerby, John Lattyak, Bruce Bowman, Pat Bernal, Wayne Young, Steve McLaughlin, Glenn Larnerd, Paul Tufano, Mark Posson, Randy Shipes, and Janet Matey-Thomas; and to my former staff members at IBM-San Jose: Fidele Alcorn, Tony Castillo, Dan Dalton, Marian Duncan, Sylvia Flores, Hugh Monini, Aviva Wada, and Andrea Zenos.

Phil Mitchell

ISO 14000
Guide

PART 1

What Is ISO 14000 and Why Is It Important?

1

What Is ISO 14000?

It is the view of the vast majority of delegates to ISO, including the participants from Europe, North America, and Asia, that the worldwide acceptance of a single EMS standard will achieve improved environmental protection in a reasonable and cost-effective manner.

CHRISTOPHER BELL
Partner, Sidley & Austin, and Chairman,
Subtag on Terms and Definitions

What Is the Aim of ISO 14000?

Today, there is a fair amount of misunderstanding concerning the ISO 14000 standards. The major reason stems from misinterpretation of what the standards aim to accomplish. Many experts in the environmental field—including environmental managers, consultants, and government officials from all countries—believe these standards prescribe worldwide environmental performance. Others want the standards to dictate environmental goals for pollution prevention, technology, or other desirable environmental outcomes. The ISO 14000 standards do none of these things. Rather, they lay out tools and systems for the management of numerous environmental obligations and the conduct of product evaluations, without prescribing what goals an organization must achieve. The ISO 14000 series, taken as a whole, aims to provide guidance for developing a comprehensive approach to environmental management and for standardizing some key environmental tools of analysis, such as labeling and life cycle assessment.

Some would contend that what ISO 14000 offers does not go far enough—that specific environmental goals are necessary in order to mitigate pollution and other environmental problems. In fact, the ISO 14000 standards can have (and assuredly will have) a significant effect on the overall environmental state of the planet. As organizations around the world begin to follow the ISO 14000 guidelines, particularly the standard for the environmental management system (EMS),* the result will "raise the floor" on overall environmental management and performance. This concept is depicted in Fig. 1-1. "Raising the floor" is a very worthy objective in light of the fact that some countries are not yet in a position to meet technology-intensive, performance-oriented environmental goals. Yet, by promoting and implementing environmental management in organizations, ISO 14000 will play a part in global environmental progress, eventually allowing all countries to catch up to those that have had environmental issues at the forefront of policy, initiatives, technology, and regulation for over two decades.

Areas of Confusion Associated with the ISO 14000 Standards

Registration

The ISO 14000 standards, in their entirety, cover a wide range of subjects. These include environmental management, environmental auditing, life cycle assessment, environmental labeling, environmental performance, and others. In all, 17 ISO 14000 documents were completed or in progress by late 1995. All the standards have value and, in many ways, are interconnected.

Because of this wide breadth of coverage, there is the potential for confusion about what is required and what is not required for ISO 14000 registration. The answer is surprisingly simple. To become registered under ISO 14000, an organization need only show its conformance to the EMS document—ISO 14001.

---

*A list of acronyms relating to ISO standards appears in Appendix A.

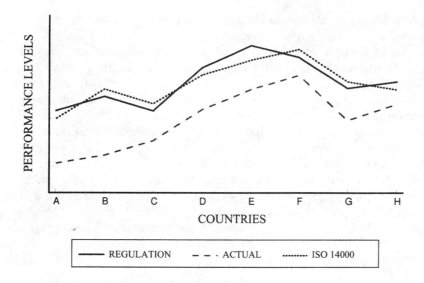

- Regulation (Solid Line): Depicts hypothetical performance levels required by regulation in a number of countries.

- Actual (Dashed Line): Depicts actual performance attained by the regulated community in those countries.

- ISO 14000 (Dotted Line): Depicts the expected performance levels if most organizations in those countries implement ISO 14000.

  Note: This illustration is not based on real data. It is only meant to show that ISO 14000 will not equalize environmental performance among countries. It does, however, have the potential to raise performance to that which is required in each country.

Figure 1-1. Performance improvement with ISO 14000.

Implementing the Standard

Views on implementation have also varied widely, with some considered decidedly off the mark. For instance, one view assumes that there is only one way to meet the elements of the EMS standard. This is not the case. The elements defined in ISO 14001 delineate a comprehensive EMS, and although they are all required to be in place, the document is flexible in its approach to implementation. Thus, there can be many customized approaches to implementing the standard, any of which may be adequate for ISO 14001 registration. The standard is meant to be applicable to large, medium-size, and small organizations in both developed and developing countries.

The EMS can be customized to suit any organization's needs. Its value is in fostering a commitment to environmental progress and to continual improvement of the organization's ability to meet its environmental responsibilities and obligations. The EMS allows organizations to meet these environmental obligations consistently and reliably, much as they meet their manufacturing or financial obligations. Enterprises in developing countries that, to date, have not made as much environmental progress as those in developed countries will not be penalized. The ISO 14001 specification does not require that an EMS be at a specific stage of advancement; rather, it requires that each element defined in the standard be part of the organization's EMS and that the EMS be integrated with the organization's other management activities.

Background Pertaining to ISO 14000

What Is the International Organization for Standardization?

The International Organization for Standardization (ISO) got its start just after World War II. ISO is a nongovernmental, international organization based in Geneva, with over 100 member bodies, or countries. It is not affiliated with the United Nations, or with any European organization, as some mistakenly assume.

Countries are represented in ISO by designated authorities within those countries. For instance, the United States is represented by the American National Standards Institute (ANSI), which is a private-sector organization. Government organizations—such as the U.S. Environmental Protection Agency (EPA), Occupational Safety and Health Administration (OSHA), and Department of Energy (DOE)—are part of ANSI, but their federal status does not confer special privileges on their membership. ANSI is both private and multisectorial. Among ISO members, ANSI is somewhat unique in this regard, since almost all other countries have chosen to be represented by organizations that are tied more closely to their governments. The various memberships in ISO, along with their designated representatives, are presented in Appendix B.

What Are ISO Standards?

The term *ISO*, which is commonly used when referring to the organization and its standards, is not an acronym, as is often assumed. *Iso* is a Greek word meaning "equal." The term is well suited to the organization, since its main focus is to provide standardization on an international level. Traditionally, ISO focused almost exclusively on product and safety standards. These technical standards have been of great value over the years and have enhanced international commerce, product uniformity, and interconnectivity.

All the standards that ISO develops are voluntary, consensus, private-sector standards. Since ISO is nongovernment, it has no authority to impose its standards on any country or organization. In addition, technical experts from ISO's member bodies develop ISO standards through a process of extensive discussion, negotiation, and international consensus. The process is an open one, and the various stakeholders and interested parties are usually well represented. Although the standards are developed for the private sector and are meant to be voluntary, government bodies may elect to convert an ISO standard to a required or legal standard. Such standards may also become conditions of doing business in commercial transactions so that parties may no longer view them as strictly voluntary. Finally, the advent of ISO 14000 environmental management standards appears to move ISO much closer to a public-sector field that it arguably has no charter to undertake.

The work of ISO is performed in technical committees established by ISO's Technical Management Board (TMB). Each technical committee receives a scope of work from TMB, and experts from the member countries then come together to begin working toward a common goal, which is the development of an ISO standard.

The Creation of ISO 14000 Standards

ISO's Development of Organizational Management Standards. During the 1980s, ISO embarked on a task to standardize one aspect of organizational management—quality management. This was the first time that ISO had ventured to create standards that were not essentially technically based and/or scientifically based. Technical Committee (TC) 176 was given the challenge to develop these quality management standards, and began work on crafting what were to become some of the most popular and successful standards in ISO's history.

The resulting standards—the ISO 9000 series—were completed in 1987. These standards have been adopted and recognized worldwide as adding value to organizations' quality management programs. Indeed, in some regions registration to the standard has become a requirement for trade.

With the success of the ISO 9000 series, ISO became confident in its ability to develop other organizational standards. By the late 1980s, riding on the reputation of the ISO 9000 standards, ISO was at the forefront of standards development. Even so, a decision to draft standards for a controversial subject in the public-sector arena, such as the environment, was not necessarily to be the next step for the traditionally conservative ISO. Additional factors beyond success with its ISO 9000 standards came into play in ISO's decision to develop environmental management standards.

International Environmental Focus.   During the same period that ISO was enjoying renown with its quality standards, much was happening internationally in the environmental arena. Ozone depletion, global warming, deforestation, and other environmental issues were making front-page news around the world, and were being viewed as global problems. Representatives from concerned countries met in Montreal in 1987 to work out agreements on banning the production of ozone-depleting chemicals. The reduction of biological diversity had also gained international focus, and a number of proposals addressing this issue were being circulated. Truly, there was an international desire for better environmental care.

Another factor that became obvious at this time was the absence of a universal indicator to assess an organization's good-faith effort to achieve reliable and consistent environmental protection. This type of indicator, which was eventually to take the form of the ISO 14001 standard, can be employed for assessment when it is coupled with an independent third-party evaluation of conformance that confirms an organization's commitment to comply with applicable country regulations, to assess significant impacts of its activities, and to develop or improve its EMS.

The early 1990s also witnessed the advent of national and regional environmental standards that could negatively impact trade. These standards were proliferating in areas such as labeling, environmental management, and life cycle assessment. In general, these standards were inconsistent with one another, and had the potential to cause serious marketplace bias between nations. Further, the inconsistencies created major harmonization problems for international enterprises. Generally, nonuniform product labeling and product assessments were yielding divergent results for the same or similar products—a situation which led, at best, to confusion and, at worst, to marketplace discrimination.

Thus, with a history of success based on ISO 9000 and with environmental issues gaining prominence, it is not surprising that ISO began to consider the environmental arena. The actual trigger, however, came in 1991. That year, the United Nations (UN) announced its Conference on Environment and Development (UNCED), to be held in June 1992 in Rio de Janeiro. In anticipation, conference representatives approached leaders of ISO's Central Secretariat to request their participation at Rio. They specifically asked that ISO make a commitment at the UNCED to create international environmental standards.

In mid-1991, on the basis of this request, ISO called on its members for volunteer advisers (some 25 countries responded) and formed an advisory group named the Strategic Advisory Group on the Environment (SAGE). SAGE decided by mid-1992 that it was appropriate for ISO to develop standards on environmental management, and this decision was made public at the UNCED. By January 1993, TC 207 was empaneled by TMB to develop environmental management systems and tools in a number of environmental areas.

SAGE's Part in the ISO 14000 Process

The mandate given to SAGE in 1991 was to assess the need for international standards in environmental management and to recommend an overall strategic plan in conjunction with the International Electrotechnical Commission (IEC). With respect to the first task, SAGE was asked to consider whether international environmental standards could serve to:

- Promote a common approach to environmental management similar to that for quality management
- Enhance organizations' abilities to attain and measure improvements in environmental performance
- Facilitate trade and remove trade barriers

Soon after its formation, SAGE broke into six subgroups, with each subgroup under the leadership of an individual country representative. The subgroups were to look at the mission defined by ISO from six different disciplines. The subgroup disciplines and corresponding lead countries included:

- Environmental management system—United Kingdom
- Environmental auditing—Netherlands
- Environmental labeling—Canada
- Environmental performance (later termed environmental performance evaluation)—United States
- Life cycle analysis (later termed life cycle assessment)—United States
- Environmental aspects in product standards—Germany

Although all six areas were considered important, the sixth subject, environmental aspects in product standards, was unique in that it was to look into the possibility of creating a guide for use by all product-standards writers. This guide would be used by writers of all standards under the ISO umbrella—not just writers of environmental standards. The purpose of the guide, as envisioned, was to help standards writers understand environmental implications associated with the standards they create. Armed with that understanding, writers would be encouraged to avoid environmentally negative attributes or specifications and select those that made the product a better environmental performer.

Although the ISO mandate clearly specified that SAGE was to evaluate only the need for standards, the subgroups of SAGE deviated from this mission and, instead, began writing standards in the various areas. Some members of SAGE had already concluded that the standards were needed, so they proceeded to develop them despite the lack of authority to do so. Further, the joint ISO/IEC strategic plan never came to fruition. IEC withdrew its support from SAGE during the process and, ultimately, was not included as part of ISO's TC 207.

The United States Takes a Position. The U.S. delegation to SAGE objected on several grounds to the subcommittee's premature drafting of standards. First, the United States felt that the SAGE committee had been called together to do advisory work and that the group should have spent its time, as it was charged to do, making a strong case for the need and desirability of ISO environmental standards. Second, the delegates to SAGE were not expected to be technical experts, and the United States insisted that ISO procedure required the involvement of experts in the writing of standards. Additionally, some countries submitted their own standards as drafts for consideration. For instance, Great Britain presented its British Standard 7750 as a draft for the environmental management system standard, with

Ireland, South Africa, and France submitting drafts of their own. The United States found draft submissions particularly objectionable, since ISO standards are required to be consensus standards, and some countries were being allowed to influence the outcome prematurely. Finally, the U.S. representatives felt that most SAGE delegates (including those from their own country) could not dedicate the time at this point in the process to produce quality standards, since their commitment was to participate only in an advisory role and not as technical experts.

The SAGE officials took no action along the lines of the U.S. objections. They felt that there was no harm in creating draft standards that could then be handed over to the technical committee. In fact, the view was that "premature" standards would give the technical committee a head start on its work. If anything needed to be changed in the drafts, the SAGE officials argued, the technical committee would have the power to make those changes.

While it was true that the technical committee had the authority to change the standards later, the countries that did not have standards in draft or completed form were put at a temporary disadvantage. Concepts from preexisting standards were embedded into the early drafts. The United States was effectively outvoted and the creation of premature standards was allowed to go forward.

Consequences for the U.S. Delegation and TC 207.   Because the United States took a stand against the creation of premature standards during the SAGE process, other countries (particularly the European countries that were bringing their standards into the process early) concluded that the United States was not in favor of creating ISO environmental standards. In fact, the United States had decided early in the SAGE process that international ISO standards in some areas were desirable for many reasons, most particularly for facilitating international trade. Regrettably, the mistaken view persisted throughout the SAGE process and never entirely dissipated during the standards-writing process under TC 207, although tensions did ease with time.

This polarization between the U.S. delegates and their European counterparts resulted in Canada being named as Secretariat of TC 207. Canada was seen as "neutral," and, although the United States had the experience and technical knowledge to lead TC 207, it agreed that Canada provided a more conciliatory locus for this undertaking. Ironically, because of persistent European misunderstanding of U.S. intentions, the United States was nearly left out of all leadership positions in TC 207. This would have been an unmerited snub for Americans and a significant loss for TC 207, since it would have reduced the level of U.S. involvement and input in the TC 207 process.

Ultimately, the United States was granted leadership of the Environmental Performance Evaluation Subcommittee. Happily, the record of U.S. contribution speaks for itself—no other country has had as much input and participation in the evolution of the ISO 14000 standards.

SAGE concluded its work at the end of 1992. Although the SAGE report did not meet the ISO mandate adequately, it did include language indicating that a technical committee should be empaneled to proceed with drafting environmental management standards. As anticipated, a number of premature concepts were embedded in the final SAGE drafts. Admittedly, the drafts helped launch TC 207 with great momentum. It was also immediately obvious that some of the premature work had to be substantially revised.

The Work of TC 207

Scope of Work Defined by the TMB.   The scope of work that was agreed to for TC 207 reads as follows:

> ISO/TC 207—Environmental Management Scope: Standardization in the field of environmental management, tools, and systems.
>   Excluded:
> - test methods for pollutants which are the responsibility of ISO/TC 146 (Air Quality), ISO/TC 147 (Water Quality), ISO/TC 190 (Soil Quality) and ISO/TC 43 (Acoustics).
> - setting limit values regarding pollutants or effluents.
> - setting environmental performance levels.
> - standardization of products.
>
>   *Note:* The Technical Committee 207 for environmental management will have close cooperation with ISO Technical Committee 176* in the field of environmental systems and audits.

This statement on the scope of work is extremely important because it establishes the boundaries for TC 207. The scope of work came out of ISO's TMB, and it is clear that TMB understood the mandate of TC 207 in terms of what the committee would and would not address.

In particular, these environmental management standards would not address the same environmental issues that were already addressed by regulatory bodies. ISO, as an international group that develops private-sector, consensus, voluntary standards, does not have the authority, experience, or expertise to establish environmental limit values,

---

*TC 176 was responsible for creating the ISO 9000 standards.

pollutant levels, technology requirements, and environmental characteristics of products. The scope of work reflects this charter and reaffirms the private-sector/public-sector divide of ISO standards. Environmental management standards that are to be voluntarily implemented by organizations are private-sector standards and totally within the purview of ISO standardization. Environmental performance, emission, or technology standards have traditionally been considered public-sector standards within the exclusive purview of government authority. The mandate given to TC 207 clearly denotes this distinction of what is included and what is excluded in its official scope of work.

The Concept of Process Standards. In its scope of work for TC 207, TMB integrated a key concept from ISO 9000 quality management standards. That concept was that management standards were process standards and, as such, were not to specify end goals.

As is now well known, the ISO 9000 standards do not address the quality of the product that is produced by an organization. Instead, the ISO 9000 standards address the quality of the process that the organization uses to create a product. It is assumed that the quality of the product is negotiated between the customer and the organization. What the customer then expects (once the quality of the product has been established) is some kind of assurance that the process that makes the product is a reliable one that can consistently turn out products of specified quality. Thus, the ISO 9000 standards focus on the management process, which in turn yields consistency of the products produced.

Some critics of the ISO 9000 standards maintain that specifying a quality process is not sufficient. They argue that ISO standards should require an organization to improve the quality of its products, or, at least, should provide limits for what is or is not acceptable in terms of product quality. Without these coercive measures or limits, critics maintain, the ISO 9000 standards have marginal usefulness. The authors contend that this is not the case at all. Customers want consistency in quality, and the ISO 9000 process provides a framework for ensuring consistency. If the established quality of the product is not acceptable, that is an issue separate from ISO 9000, and should be negotiated between the organization and the customer.

Similar to the ISO 9000 approach, the scope of work for ISO 14000 environmental standards expects TC 207 to develop management, or process, standards. These will allow organizations, anywhere in the word, to consistently meet their environmental obligations on all fronts—regulatory, community, employee, and stockholder. This consistency in meeting environmental obligations is synonymous with the consistency of producing the same quality product under ISO 9000. Like ISO 9000, these standards are intended not to set performance goals, benchmarks, or limits, but to specify the elements of a system that aims to achieve a consistent and reliable process to consistently meet environmental obligations.

Misconceptions About the Scope of Work Given to TC 207. Regardless of the clear intent and unambiguous wording of the mandate given to TC 207, a surprising (and sometimes exasperating) number of delegates from various countries—including the United States— have consistently failed to understand the limits on the TC 207 scope of work. These individuals, who sometimes comprise entire country delegations, evidently do not understand the difference between an end goal and a process. They persist in trying to incorporate performance objectives into the standards, either by specifying required levels of technology or by demanding direct, continuous environmental performance improvements rather than process mechanisms.

From the U.S. point of view, a performance focus is both inconsistent with what TC 207 is authorized to do and a potential problem, as it encroaches on the authority of governments and regulatory bodies to establish performance, end-goal requirements. ISO normally limits its reach to the publication of private-sector standards. Environmental performance standards traditionally arise through the political process which gives them shape and, ultimately, life through government pronouncements. If the member countries of ISO had chosen to transfer environmental authority to ISO, then TC 207 might have sought to elaborate performance standards. With one exception, no such broad transfer of authority has occurred, nor is it likely to occur any time soon. The exception exists in the European Union (EU), which has effectively transferred such authority to the European Committee for Standardization (CEN). This state of affairs does not exist outside of Europe—certainly not in the United States. The difference in government approach to standards between Europe and the rest of the world was the principal cause of all difficulties experienced in drafting the ISO 14000 series. Fortunately, there was a broad consensus outside the EU that ISO standards should not encroach on or replace national legislation and regulation that specifies the level of environmental achievement a sovereign nation wishes to attain.

The Structure of TC 207

TC 207 is an international committee made up of the member bodies of ISO. Thus, it is possible to have over 100 delegations working on the ISO 14000 standards. In actuality, 50 countries have indicated a desire to participate in the work of TC 207, and of this group, only about half are actively contributing. A list of participating, observer, and liaison members of TC 207 is shown in Fig. 1-2.

As mentioned previously, Canada holds the Secretariat position for TC 207. As Secretariat, Canada is responsible for overall organization of the group, document distribution, and other administrative duties that

---

TC 207 Members and Liaison Organizations

Participating Members: Argentina, Australia, Austria, Belgium, Brazil, Canada, Chile, China, Colombia, Cuba, Czech Republic, Denmark, Ecuador, Finland, France, Germany, India, Indonesia, Ireland, Israel, Italy, Jamaica, Japan, Republic of Korea, Malaysia, Mauritius, Mexico, Mongolia, Netherlands, New Zealand, Norway, Philippines, Romania, Russian Federation, Singapore, South Africa, Spain, Sweden, Switzerland, Tanzania, Thailand, Trinidad and Tobago, Turkey, Ukraine, United Kingdom, United States, Uruguay, Venezuela, Zimbabwe

Observing Members: Algeria, Barbados, Croatia, Egypt, Estonia, Greece, Hong Kong, Iceland, Libya, Lithuania, Poland, Portugal, Slovakia, Slovania, Sri Lanka, Viet Nam, Yugoslavia

Internal Liaison Organizations: ISO/TC 61 on Plastics, ISO/TC 176 on Quality Management and Quality Assurance, ISO/TC 203 on Technical Energy Systems, IEC/TC 75

External Liaison Organizations: Consumers International, Environmental Defense Fund, European Chemical Industry Council, European Environmental Bureau, Friends of the Earth International, Industrial Minerals Association, International Chamber of Commerce, The International Council on Metals and the Environment, International Electrotechnical Commission, International Federation of Organic Agriculture Movement, International Institute for Sustainable Development, International Iron and Steel Institute, International Network for Environmental Management, International Primary Aluminum Institute, International Trade Centre, Organisation for Economic Co-operation & Development, United Nations Conference on Trade and Development, United Nations Environment Program, World Wide Fund for Nature

---

Figure 1-2. TC 207 membership, mid-1995.

ensure that the work of TC 207 proceeds according to plan. The member body of ISO from Canada is the Standards Council of Canada (SCC). The SCC has delegated the responsibility of Secretariat for TC 207 to the Canadian Standards Association (CSA). Part of the Secretariat's duties entails the selection of the chairman of TC 207. Dr. George Connell, past president of the University of Toronto, serves as chairman.

The Subcommittees of TC 207. The initial work of TC 207 included the same subjects that were reviewed by SAGE, plus a new topic—harmonization of terms and definitions. In order to carry out the work, six subcommittees and one working group were formed to address each of the areas. Various country members were assigned leadership roles for each of the six subcommittees and the working group. These subcommittees and the working group are depicted in Fig. 1-3.

It should be emphasized that the role of Secretariat is administrative only. The standards developed in each subcommittee are international consensus standards, and all participating countries have the right to be involved equally in developing those standards.

Under SAGE, the subcommittee led by the United States had been designated as Environmental Performance. When it was confirmed that ISO 14000 would not be prescribing environmental performance standards on a worldwide basis, the scope of work for Subcommittee 4 of TC 207 was changed to evaluation of environmental performance, and thus the subcommittee's name became Environmental Performance Evaluation.

Because the scope of work of TC 207 is defined as the standardization of environmental management systems and tools, anything that fits under that category can be addressed by the committee. Thus, the original list of seven work items is not necessarily the final list. Additional work items that are being proposed to the committee include design for the environment, environmental procurement, occupational health and safety, and others.

The U.S. Technical Advisory Group. ANSI is the member body of ISO from the United States. Consistent with its own procedures, ANSI formed a technical advisory group (TAG) to represent U.S. interests and create U.S. input into TC 207. The TAG is the vehicle for obtaining consensus for U.S. positions, as well as for creating documents that are then submitted to the various TC 207 subcommittees. Delegates of the TAG negotiate positions for the United States at international working-group, subcommittee, and committee meetings.

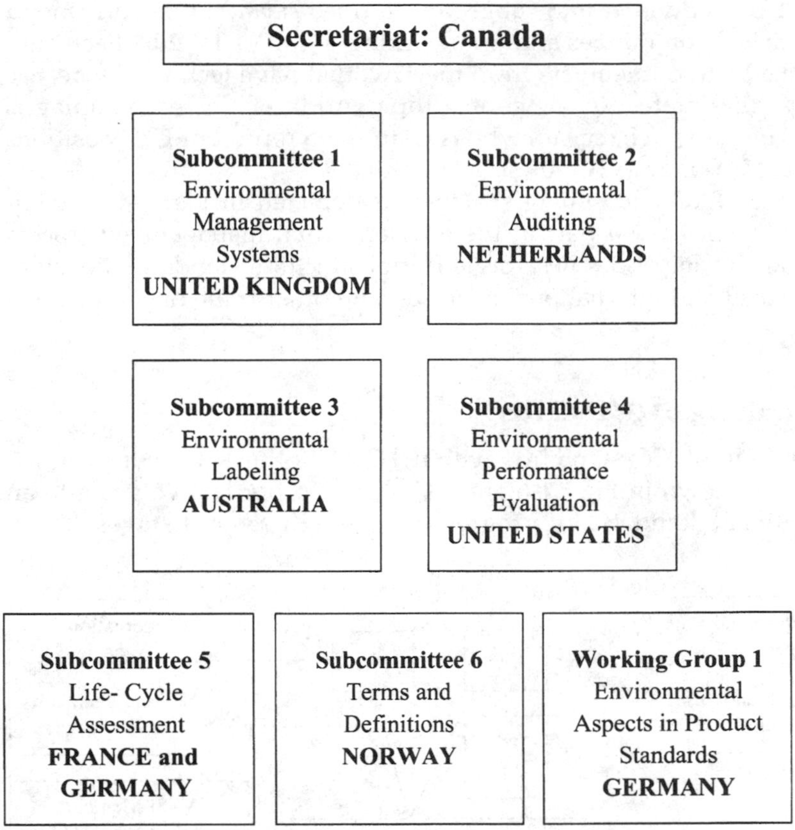

Secretariat: Canada

Subcommittee 1 Environmental Management Systems **UNITED KINGDOM**	Subcommittee 2 Environmental Auditing **NETHERLANDS**
Subcommittee 3 Environmental Labeling **AUSTRALIA**	**Subcommittee 4** Environmental Performance Evaluation **UNITED STATES**

Subcommittee 5 Life- Cycle Assessment **FRANCE and GERMANY**	Subcommittee 6 Terms and Definitions **NORWAY**	Working Group 1 Environmental Aspects in Product Standards **GERMANY**

Figure 1-3. ISO/TC 207 structure.

ANSI delegated the administration of the TAG to two organizations, the American Society of Testing and Materials (ASTM) and the American Society for Quality Control (ASQC). These organizations were given responsibility to administer the work of the TAG and its various subgroups. The TAG has over 470 members (as of late 1995) that represent a variety of different interests, including industry, business associations, environmental groups, consultants, academicians, and government. Government representation includes EPA, DOE, and the Department of Commerce. The TAG makes continuous efforts to expand its membership to provide a balance among all interested parties and stakeholders in the environmental arena. The TAG elects a chairman who represents the U.S. positions at the international meetings of TC 207.

The TAG is divided into subgroups, termed *subtags,* that correspond to the six subcommittees and one working group of TC 207. Each subtag is made up of members from the TAG that have technical expertise or a specific interest in the subtag topic. Additionally, each subtag is headed by it own chairman, who is charged to represent U.S. positions in the corresponding TC 207 subcommittee.

The U.S. TAG and subtag structure is depicted in Fig. 1-4. Organizations which have a part in the environmental management process are shown in Fig. 1-5. These organizations include standards developers, national and international agencies, and other trade or professional entities.

The Drafting of Standards

The drafting of ISO standards within TC 207 follows a predetermined course, as shown in Figs. 1-6 and 1-7. A new proposal evolves into an international standard by progressing through several stages. These

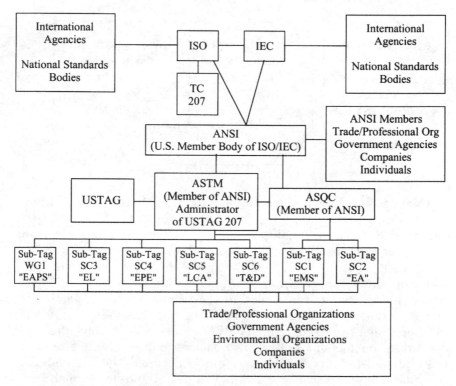

Figure 1-4. U.S. TAG and subtag structure.

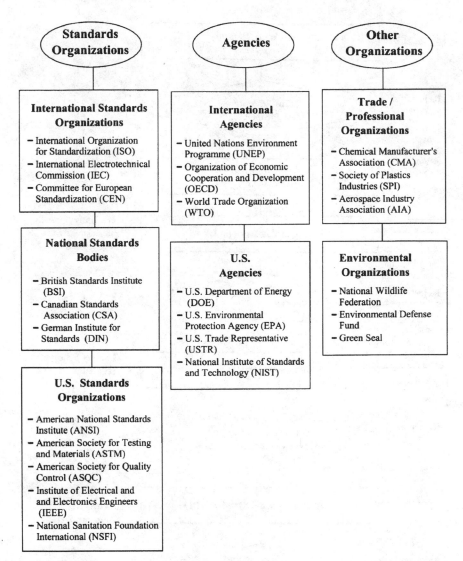

Figure 1-5. Organizations that have a part in the environmental management process.

include proposal review, working draft, committee draft, draft international standard, and, ultimately, international standard.

A new work item proposal must be approved by a majority of participating members of the TC 207 subcommittee. An additional requirement is that five participating members of the subcommittee must be willing to work on the item. As work on the new item pro-

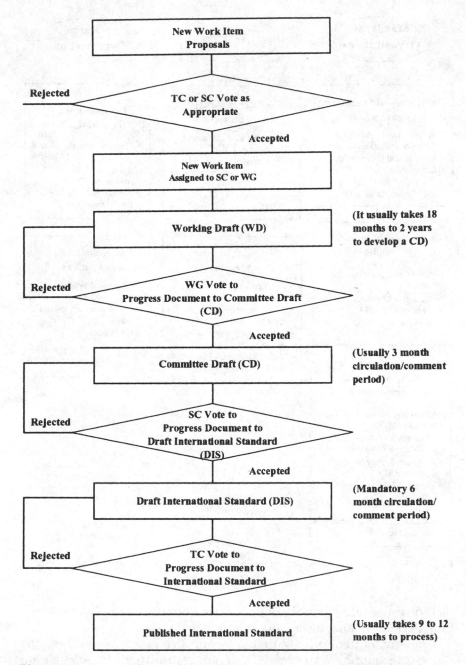

Figure 1-6. Protocol for new ISO standards. (SOURCE: USQC [1987].)

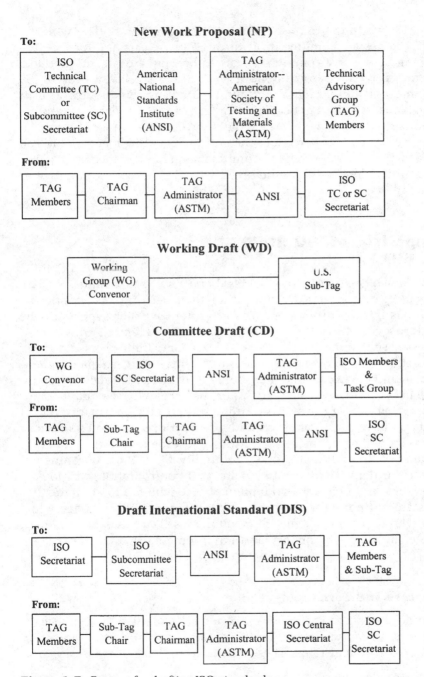

Figure 1-7. Process for drafting ISO standards.

gresses, a working group formulates a working draft. The working draft becomes a committee draft through consensus approval by the subcommittee. For a committee draft to become a draft international standard, it must be formally approved, through a ballot process, by two-thirds of the participating members voting. Finally, for a draft international standard to become an international standard, two-thirds of the participating members of the subcommittee must approve it, and not more than 25 percent of the entire ISO voting membership can vote against it. The entire ISO voting membership includes all member bodies of ISO, not just those involved with the technical committee or subcommittee.

Comparison of ISO 9000 and ISO 14000

As explained previously, the ISO 9000 and ISO 14000 standards share the goal of developing process rather than performance standards. There has been additional effort to harmonize other aspects of the standards. Structure, terminology, and other elements have been addressed so that the standards are at least compatible.

Nonetheless, there are some major differences between quality management and environmental management that impede total correspondence between the two standards. For instance, whereas quality standards affect an organization and its customers, environmental standards have a greater reach and affect an organization's relationship to its neighbors, nearby creatures and ecologies and, ultimately, humankind. In addition, unlike the quality field, the environmental arena is burdened with a bitter history of confrontation, ideological battle lines, and political exploitation. Those who fail to attain quality levels are not normally subject to civil and criminal sanctions, while those who transgress the environmental laws clearly are. So, although some elements of the quality and environmental management standards can be similar, of necessity others will be dissimilar.

Structure and Terminology

Three documents in the ISO 9000 series are classified as *requirements documents,* since they lay down the requirements for an organization that wants to be registered. These standards in the ISO 9000 series are:

- ISO 9001—model for quality assurance in design/development, production, installation, and servicing

- ISO 9002—model for quality assurance in production and installation
- ISO 9003—model for quality assurance in final inspection and test

In the ISO 14000 series, the equivalent to these three standards is ISO 14001. This standard is called a *specification document,* since it provides the specifications for an EMS. An organization must meet these specifications if it wants to become registered. Both the ISO 9000 and the ISO 14000 documents provide the same type of architecture, even though the terminology differs. In effect, requirement and specification documents are synonymous in their intent and effect.

In addition, both the ISO 9000 and the ISO 14000 series include a *guidance document* for developing and implementing their respective management systems. In the ISO 9000 series, the guidance document is ISO 9004, and in the ISO 14000 series, it is ISO 14004.

Some components of the ISO 14000 series have no parallel in the ISO 9000 standards. These include environmental tools such as environmental labeling and life cycle assessment, as well the environmental performance evaluation guideline document. While the numbering system has some parallels, a one-to-one correspondence between document numbers or content does not exist. A detailed comparison of ISO 9000 and ISO 14000, showing the similarities and differences between ISO 9001 and ISO 14001, is presented in Table 1-1.

Efforts have been made to harmonize terminology between ISO 9000 and ISO 14000. However, there are numerous instances—beginning with the use of the term *requirement* versus *specification,* as described in the previous section—in which the terminology is not consistent. At present, alignment of ISO 9001 and ISO 14001 is being pursued at a high level (i.e., for the major requirements and structures of the standards). However, it is impractical to think that alignment of the two documents can take place in the short term, particularly since new editions of ISO 9000 were published in 1994; target dates for alignment are 1999/2000.

Auditing

Auditing is part of both the ISO 9000 and the ISO 14000 standards. There are some similarities between the two corresponding documents (such as the consistent use of the verb *should*), but, in general, the documents differ markedly.

The ISO 9000 auditing documents are ISO 10011-1, which is the guideline for auditing quality systems; ISO 10011-2, which includes

Table 1-1. Comparison of ISO 9000 and ISO 14000

	ISO 9000	ISO 14000
Aims	Provides to supplier organizations a means for demonstrating to customer organizations the achievement of requirements for quality; enhances the achievement of a supplier organization in providing overall performance in relation to objectives for quality.	Provides organizations with the elements of an environmental management system; provides assistance to organizations considering the implementation or improvement of an environmental management system, including advice on enhancing such a system to meet environmental performance expectations.
Structure	Mixture of management activities, process requirements, and verification requirements; separate guidance standard.	Adheres to "plan-do-check-act" type of business model; separate guidance standard.
Contents	Both ISO 9001 and ISO 14001 include the elements of management commitment and responsibility, management system documentation, document control, operational control, training, monitoring and measurement, nonconformance and corrective action, records, and audits	
	ISO 9001 includes discrete elements of quality planning, product identification and traceability, and statistical techniques.	ISO 14001 includes discrete elements of environmental aspects, legal requirements, objectives and targets, environmental management program, communications, and emergency preparedness and response.

the qualification criteria for quality system assessors; and ISO 10011-3, which is the procedure for the management of assessment programs. The guidelines apply to internal quality system assessments that are required by the standard, as well as to external and extrinsic quality system assessments. In addition, ISO Guide 48 specifies guidelines for third-party assessment and registration of a supplier's quality system. The ISO 14000 auditing documents are ISO 14010, which contains the guidelines for general principals for environmental auditing; ISO 14011, which specifies the procedures for auditing an EMS; and ISO 14012, which includes the qualification criteria for environmental auditors of the EMS.

In general, the ISO 9000 auditing documents allow a degree of subjectivity on the part of the auditor to ascertain the substance of the quality management system. This is not the case with the ISO 14000 auditing documents, which tend to limit the auditor's subjectivity. The auditor of the EMS must use objective criteria when conducting the audit, and must compare the auditee's EMS against these predetermined criteria to assess conformance. Essentially, auditing areas that do not relate to the management of a process—such as an organization's environmental performance, use of environmental technologies, or compliance to regulations—is not expected under ISO 14000, except insofar as assessments of these elements serve to indicate whether the management system is working toward meeting the policy and objectives of the organization.

Because of the inherent differences between quality management and environmental management, there has been little effort to harmonize the amount of auditing subjectivity allowed in ISO 14000 with that of ISO 9000. Many participants in the ISO 14000 process feel that environmental auditing is a very sensitive area from a legal standpoint; thus, members of TC 207 have attempted to limit the scope of the environmental audits. Quality management does not have the same liabilities that environmental management does, so the approach to auditing is less controversial and can allow the auditor more latitude.

Documentation

One of the strongest aspects of the ISO 9000 management model is the requirement for sound, comprehensive, controlled documentation. This model, which requires documentation for all major elements of the quality system, is also included in the ISO 14000 standards. Records management, document control, documented procedures, and training records must all be part of the EMS.

British Standard 7750 and Europe's Eco-Management and Audit Scheme Regulation

Other environmental standards which have some of the same elements as ISO 14000, and which may be considered as rivaling ISO 14000, are British Standard (BS) 7750 and Europe's Eco-Management and Audit Scheme Regulation (EMAS). A comparison of the three standards is presented in Table 1-2.

Table 1-2. Comparison of ISO 14001, British Standard (BS) 7750, and European Union's Eco-Management and Audit Scheme Regulation (EMAS)

	ISO 14001	BS 7750	EMAS
Type of standard	Voluntary, consensus, private-sector standard.	National, voluntary standard	European Union regulation
Applicability	Can apply to the whole organization or part of an organization; applicable to an organization's activities, products, and services in any sector; applicable to nonindustrial organizations such as local government agencies and nonprofit organizations.	Can apply in the United Kingdom and other developed countries; can apply to the whole organization or part of an organization; applicable to all activities and sectors; applicable to nonindustrial organizations such as local government agencies and nonprofit organizations.	Applies to European Union; applies to individual facilities; applies to site-specific industrial activities.
Focus	Focuses on environmental management system; indirect link to environmental improvements.	Focuses on environmental management system, with environmental improvements emerging from the system.	Focuses on environmental performance improvements at a site and the provision of communication of improvements to the public.
Initial environmental review	Suggested in annex, but not required in standard.	Suggested but not specified in standard.	Required in regulation.
Policy commitment	Policy commitment to continual improvement of the environmental management system and to prevention of pollution; policy commitment to compliance with applicable environmental legislation and voluntary commitments.	Policy commitment to continuous improvement of environmental performance.	Policy commitment to continuous improvement of environmental performance and compliance with applicable environmental legislation.

Audits	Audits of the environmental management system are required; monitoring and measuring of key environmental characteristics are required; frequency of audits is not specified.	Audits of the environmental management system are required; audits for compliance or environmental performance are not required; frequency of audits is not specified.	Audits of environmental management systems, processes, data, and environmental performance are required; audits are required at least every 3 years.
Public communication	Only environmental policy must be made public; other external communications must be considered, but what is communicated is left to management.	Only environmental policy must be made public; other external communications must be considered, but what is communicated is left to management.	A description of the environmental policy, program, and management system must be made available to the public; a public environmental statement and annual simplified statement including factual data are required.

BS 7750 was actually in progress in 1991 when SAGE was deliberating to advise ISO about the need and desirability of developing international environmental management standards. In fact, the original draft standard of BS 7750 became the model for the work of SAGE and, subsequently, considerably influenced ISO 14001, the environmental management specification. This was, in fact, the second time that the British had led with a draft standard on a topic for which ISO later created international standards. The first was BS 5750, a quality management standard which served as the model for the ISO 9000 standards. In addition, the British have developed a draft guide for occupational health and safety, BS 8750. At this point it is not known if ISO will elect to draft standards for this area, which, like the environmental area, is fraught with controversy.

EMAS, developed by the EU, followed the development of BS 7750. Both BS 7750 and EMAS take a process and a performance approach to environmental management. Unlike ISO 14000, BS 7750 and EMAS are full-system standards; that is, with these systems, virtually no other regulation is needed. The standards include the key elements defined in ISO 14000—such as requirements for policy statements, setting of environmental objectives and targets, assessment of environmental impacts of activities, training, and documented environmental proedures. They go beyond these requirements, however, in that they include other specifications for continuous improvement of environmental performance, communication of goals and attainments to the public, and maintenance of comprehensive environmental registers.

Although specific pollution or effluent limits are not established in those standards, there is an explicit requirement for continuous improvement of environmental performance. This means that the organization must carefully track pollutants or effluents emitted to the environment, and must initiate methods for lowering the amounts discharged, regardless of whether it is already in full compliance with all applicable laws and regulations. Essentially, this requirement for continuous improvement constitutes a mandate on performance that transcends the scope of work given to TC 207.

In addition, the EMAS regulation also requires an organization to communicate its objectives and targets to the public and, at certain intervals, to disclose the organization's progress as well, including successes and failures in meeting these objectives and targets. Disclosure allows public intervention at two levels. First, the public can view a company's objectives and targets and ascertain which companies have set the most (and least) aggressive goals for pollution prevention; second, the public can exert pressure on those companies which have less aggressive goals or which do not meet their objectives and targets.

This approach is very similar to a compliance and enforcement tool like the Toxic Release Inventory (TRI) system in the United States. That system, known as Title III of the Superfund Amendments and Reauthorization Act (SARA) of 1986, requires among other things that companies report to EPA all releases to the environment beyond a certain threshold quantity of over 300 hazardous and toxic chemicals. Even though the overwhelming majority of the releases were permitted under federal or state law, once the information was released to the public (the first reports were made public in 1988), companies with significant discharges were pressured by neighborhood associations, environmental groups, and other public organizations to reduce emissions. The reports are published annually, and the public pressure for pollution prevention efforts has led to substantial—in some cases, dramatic—reductions in emissions. The Europeans hope EMAS will have similar success in Europe.

European delegates have argued during the standards-writing process that the ISO 14000 standards should follow the lead of BS 7750 and EMAS, and should include performance-oriented goals as well as strong communication requirements. In effect, the Europeans want the standards to serve as regulatory objectives while providing a process management system. The United States has objected strongly to this approach, since the scope of work of TC 207 has clearly indicated that the ISO 14000 standards are to be only process-type standards. As such, the ISO 14000 standards are to be complementary to national regulatory regimes. The ISO 14000 standards are not intended to replace or duplicate a country's regulatory system.

It is a fact that, in general, European environmental regulatory systems are not as strong as the environmental regulatory systems in the United States, and this is the likely reason delegates from that region want the ISO 14000 standards to serve as reinforcement (or perhaps enforcement) for each country's regulatory practices. The obvious compromise is to use ISO 14000 as a basis upon which to build additional requirements that satisfy unique national or local needs. The requirements of EMAS can simply be added to ISO 14000 to satisfy European needs. EPA is already investigating the use of ISO 14000 in the United States in just that way.

Most countries outside of Europe have gone along with the U.S. position that ISO 14000 should be limited to process-type management standards that complement a country's own laws and regulations. Most feel that a supplementary regulatory system either is not needed or is undesirable, and that the additional requirements of mandating environmental performance goals and public communication are better implemented through other means, on a country-by-country basis. In fact, it might be an unfair trade barrier to prescribe international standards for environmental performance that developing countries could not achieve.

2
What Are the Elements of ISO 14000?

*It is a major challenge to be sure that we
create a standard that has as great a value to
small countries that are developing, as to
larger nations.*

DOROTHY BOWERS
*Vice President, Environmental and Safety
Policy, Merck & Co., Inc., and Chairman,
Subcommittee 4 (Environmental Performance
Evaluation) of TC 207*

The subjects covered under ISO 14000 can be divided into two separate areas. The first deals with an organization's management and evaluation systems; the second, with environmental tools for product evaluation. This division within the ISO 14000 generic family of standards is illustrated in Fig. 2-1. As shown, organization evaluation consists of three subsystems that include the environmental management system, environmental auditing, and environmental performance evaluation. Product evaluation consists of three separate applications and includes environmental aspects in product standards, environmental labeling, and life cycle assessment. A separate effort is focused on terms and definitions to harmonize their usage among all areas and applications under ISO 14000. The various products that either are currently in progress or have been completed are presented in Figs. 2-2 and 2-3.

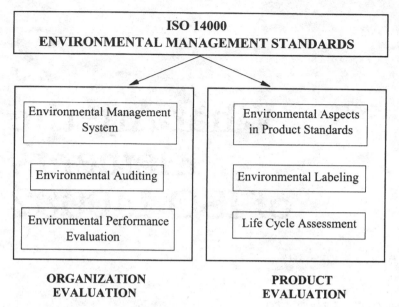

Figure 2-1. ISO 14000 family of standards.

Organization Evaluation

Environmental Management System

The ISO 14001 document entitled "Environmental Management Systems—Specification with Guidance for Use" is arguably the most consequential in the ISO 14000 series. This standard lays out the elements of the environmental management system (EMS) that organizations are required to conform to if they seek registration, or certification, after passing an independent third-party audit from an accredited registrar.

Because ISO 14001 is so important, over 75 percent of the participating and observing members of TC 207 are also either participating or corresponding members of Subcommittee 1, which is responsible for drafting the EMS standards. Member bodies of Subcommittee 1 are presented in Fig. 2-4.

The 14004 document entitled "Environmental Management Systems—Guidelines on Principles, Systems, and Supporting Techniques" provides supplementary information. This product is not designed for registration or certification, and there are a number of warnings in the document not to use it for that purpose. The warnings are directed primarily at registrars who will assess an organization's

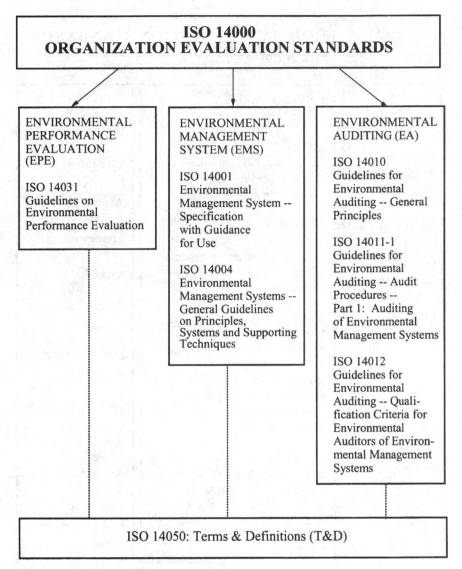

Figure 2-2. ISO 14000 organization evaluation standards.

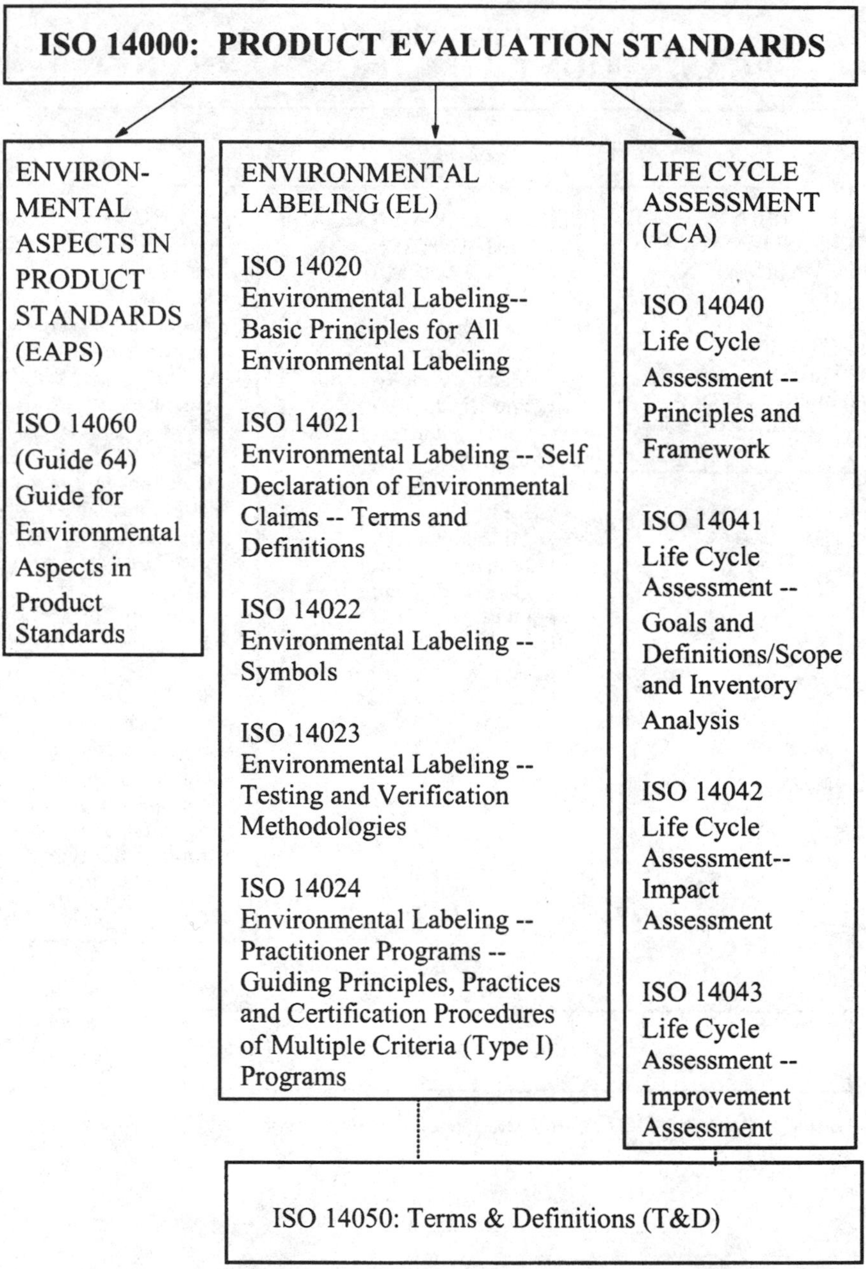

ISO 14000: PRODUCT EVALUATION STANDARDS

ENVIRON-MENTAL ASPECTS IN PRODUCT STANDARDS (EAPS)	ENVIRONMENTAL LABELING (EL)	LIFE CYCLE ASSESSMENT (LCA)
ISO 14060 (Guide 64) Guide for Environmental Aspects in Product Standards	ISO 14020 Environmental Labeling-- Basic Principles for All Environmental Labeling	ISO 14040 Life Cycle Assessment -- Principles and Framework
	ISO 14021 Environmental Labeling -- Self Declaration of Environmental Claims -- Terms and Definitions	ISO 14041 Life Cycle Assessment -- Goals and Definitions/Scope and Inventory Analysis
	ISO 14022 Environmental Labeling -- Symbols	
	ISO 14023 Environmental Labeling -- Testing and Verification Methodologies	ISO 14042 Life Cycle Assessment-- Impact Assessment
	ISO 14024 Environmental Labeling -- Practitioner Programs -- Guiding Principles, Practices and Certification Procedures of Multiple Criteria (Type I) Programs	ISO 14043 Life Cycle Assessment -- Improvement Assessment

ISO 14050: Terms & Definitions (T&D)

Figure 2-3. ISO 14000 product evaluation standards.

Subcommittee 1 Members

Participating Members: Argentina, Australia, Austria, Belgium, Brazil, Canada, Chile, Colombia, Cuba, Czech Republic, Denmark, Finland, France, Germany, India, Indonesia, Ireland, Italy, Jamaica, Japan, Republic of Korea, Malaysia, Mexico, Netherlands, New Zealand, Norway, Philippines, South Africa, Spain, Sweden, Switzerland, Tanzania, Thailand, Trinidad and Tobago, Turkey, Ukraine, United Kingdom, United States, Uruguay, Zimbabwe

Correspondent Members: Iceland, Israel, Mongolia, Poland, Portugal, Singapore, Slovakia, Sri Lanka, Yugoslavia

---

Figure 2-4. Membership in TC 207 Subcommittee 1 on environmental management systems, mid-1995.

conformance to ISO 14001. These warnings are considered necessary to prevent any expansion of the ISO 14001 requirements with guidance found in ISO 14004. ISO 14004 is meant to be used only as guidance by organizations that are just beginning to implement an EMS. For that purpose, ISO 14004 may be a useful adjunct to ISO 14001 in both developing countries and in small and medium-size enterprises (SMEs). In addition, the guidance includes some elements and suggestions that can be followed to further improve an existing EMS. These elements and suggestions are intended for organizations that have a mature, sophisticated EMS.

ISO 14001.   ISO 14001 is the management system specification document in the ISO 14000 series. It contains those elements that must be satisfied by an organization seeking registration or certification to the standard. Its function is similar to ISO 9001, 9002, and 9003 in the ISO 9000 series, which, as mentioned in Chap. 1, are called requirements documents. The terms *specification* and *requirements* are used interchangeably in ISO literature. Additionally, in the United States, the term *registration* is preferred to avoid legal warranty implications suggested by the term *certification.*

The elements detailed in ISO 14001 must be implemented, documented, and executed in such a way that an independent third-party registrar can grant and justify registration on the basis of evidence that the organization has implemented, in good faith, a viable EMS. ISO 14001 is also designed for organizations that wish to declare their conformity to the standard to second parties that are willing to accept such self-declaration without the intervention of third parties.

The major challenge for both the organization that is implementing ISO 14001 and the registrar who is auditing that organization's conformance is uniformity in interpretation of the specifications. What is meant, what is expected, and what does it take to become registered to ISO 14001? The paragraphs below provide some guidance on the specification elements, as well as clarification on potentially ambiguous or misleading terms and concepts.

EMS Structure. An environmental management system is "that part of the overall management system which includes organizational structure, planning, activities, responsibilities, practices, procedures, processes, and resources for developing, implementing, achieving, reviewing, and maintaining the environmental policy" (Section 3.5). Elements of the management system, as described in the definitions and in other places in the standard, are pictorially represented in Fig. 2-5.

As portrayed, the elements can be visualized as building blocks of a pyramid, with the core elements of management commitment and environmental policy forming the base for all the other components of the EMS. The second tier of the pyramid contains an organization's environmental goals, objectives, and targets; and the third tier is the embodiment of these goals, objectives, and targets in an environmental

Figure 2-5. Environmental management system pyramid.

management program made up of processes, practices, procedures, and lines of responsibility.

ISO 14001 calls for the establishment of one or more environmental programs to achieve the objectives and targets set by the organization. To a large extent, the suitability and effectiveness of the EMS is periodically assessed by management reviewing the progress achieved through these environmental programs. That progress is tracked by a performance evaluation subsystem which feeds its output directly into the management review process.

Another significant input into the management review process comes from the periodic EMS audits which comprise the fourth tier. The purpose of these audits is to ascertain that the EMS is being maintained and that it works the way it was intended to work. Such audits are also used to assess the compliance and management review processes themselves. Management review is the fifth tier, and is designed to determine the adequacy, suitability, and effectiveness of the EMS by management on the basis of all inputs. The final tier highlights the pinnacle goal, which is to achieve continual improvement of the EMS in order to ensure that the organization is consistently and reliably meeting its environmental obligations and protecting the environment.

When viewed as a pyramidal structure, it is easy to see that elements in the lower tiers of the EMS are critical building blocks of the system and must be in place to support the elements above. Further, continual improvement is not achievable without all aspects of the EMS in place. The EMS is designed to provide a structure and systematic approach to overall environmental management.

Definitions. The definitions in Section 3 of ISO 14001 are intended to assist users in achieving uniform interpretation and implementation of its requirements. Because of the difficult and controversial nature of the subject and the protracted debates in the development of this standard, it is expected that final interpretation of these requirements will be achieved iteratively over the next few years. The authors believe that, of itself, this does not detract from the usefulness of the standard and will not seriously compromise national conformity assessment programs. Rather than being prescriptive, the definitions serve to set the tone for the underpinnings of the standard, and provide guidance for the key elements of the environmental management system.

Application of ISO 14001. ISO 14001 has been written to be applicable to organizations of all types and sizes and to accommodate diverse geographical, social, and cultural conditions. This type of system enables an organization to establish and assess the effectiveness of procedures that set environmental policy and objectives and that achieve conformance to both. Further, the EMS allows an organization to demonstrate conformance of its policy, objectives, and procedures to others through a third-party audit or through self-declaration of conformance.

Conformance to ISO 14001 can be, of itself, an indicator of good faith and commitment to environmental protection. Implementing environmental management techniques in a systematic manner provides the opportunity for environmental improvement and consistency in meeting environmental responsibilities.

Technology Specifications.   ISO 14001 applies to all kinds of organizations, since technology requirements are not a part of the standard. In its introduction, ISO 14001 does encourage organizations to consider implementation of best available technology where appropriate and where economically viable. However, there is no requirement in 14001 to use the best available technology, or any other technology for that matter. The only technological requirement in this standard is the obligation to consider "prevention of pollution" options when designing new products or systems. Prevention of pollution is defined as "use of processes, practices, materials, or products that avoid, reduce, or control pollution, which may include recycling, treatment, process changes, control mechanisms, efficient use of resources, and material substitution. *Note:* The potential benefits of prevention of pollution include the reduction of adverse environmental impacts, improved efficiency, and cost reduction" (Section 3.13). A registrar conducting a registration audit will seek evidence that these options have been considered even if the chosen solution is one that merely controls pollution.

Environmental Performance.   Environmental performance is defined in the specification document as "measurable results of the environmental management system related to an organization's control of the environmental aspects, based on its environmental policy, objectives, and targets" (Section 3.8). Performance indicators might include adherence to environmental permit, reporting, and other regulatory requirements; adherence to administrative requirements such as labeling; adherence to training schedules; and improvements in environmental areas such as recycling, energy conservation, and pollution prevention, as outlined in the environmental objectives and targets.

Environmental performance is not considered apart from the EMS; rather, it is totally dependent on the criteria established by the management system. Societal expectations of desired performance are considered in the organization's setting of objectives and targets. These, however, are set by management and are not subject to reassessment by a third-party registrar, whose only role is to evaluate the process by which stakeholder views are taken into account when objectives and targets are set. There are no specific performance requirements in ISO

14001 beyond commitment, in the form of policy, to compliance processes for applicable legislation and regulation, to continual improvement of the EMS, and to serious attempts to prevent pollution.

Occupational Health and Safety Management. ISO 14001 does not currently include requirements for aspects of occupational health and safety (OH&S) management. An ISO process is now in progress to deliberate on and decide whether OH&S management standards are needed on a worldwide basis. For the time being, OH&S aspects can be integrated with environmental elements if the organization's preference is to integrate these related areas of activity. ISO 14001 does not preclude such inclusion. The expectation, however, is that the registration audit will not cover and may not be competent to assess elements of OH&S and should not automatically be relied on for assurances in that field.

Major Elements Contained in ISO 14001. The five major elements associated with the EMS, as outlined in ISO 14001, are depicted in Fig. 2-6. Each of these elements is discussed in detail in Part 2 of this book. The discussion includes a look at definitions contained in the specification document, clarification and interpretation of major aspects of the document, and examples of methods and techniques that can be used by an organization when implementing ISO 14001.

Annex A (Informative). ISO 14001 includes an annex which simply augments the frugal wording in the ISO 14001 specification in order to lend it more substance and conviction. The content, however, has no substantive effect—both because the annex is "informative" and therefore nonusable for registration and because there is nothing there that is not

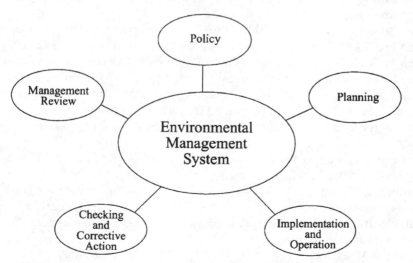

Figure 2-6. The five major elements of an EMS.

already in the specification. Still, the hope is that the additional language may help clarify some aspects of the specification for some users.

Because of the informative rather than "normative" classification of the annex, registrars are not allowed to use it to add requirements they believe are not already included in the specification. This may change someday, as future revisions of ISO 14001 may see this annex absorbed into the specification portion or it may become a part of the information guidance in ISO 14004.

Annex A was deemed unnecessary by many delegates to TC 207. It was included only for political reasons related to the European Union's (EU's) Eco-Management and Audit Scheme Regulation (EMAS), and it is hoped that the descriptive narrative within it will prove reassuring and comforting to our European colleagues. Within the EU, implementation of EMAS may eventually require that some portions of the annex be treated as mandatory.

ISO 14004

Application of ISO 14004.   ISO 14004 is "informational" and can be used by SMEs just beginning to structure an EMS, or by larger organizations trying to improve or optimize an existing system. The guidance document clearly states in its introduction that only ISO 14001 contains the requirements that may be objectively audited for certification or registration purposes or for self-declaration purposes. Thus, ISO 14004 is not to be used for registration. Instead, ISO 14004 includes examples, descriptions, and options, as well as practical advice, that will aid both in implementing or enhancing an EMS and in strengthening its integration into the overall management of the organization.

The Structure of ISO 14004.   ISO 14004 is structured to broadly mirror ISO 14001 in terms of the five major topics included—environmental policy, planning, implementation, checking and corrective action (termed "measurement and evaluation" in the guidance document), and management review. In addition, all subtopic specifications within ISO 14001 are covered in ISO 14004, as well as additional (optional) topics. A comparison of information in ISO 14001 and ISO 14004 is presented in Table 2-1.

The guidance document includes practical help on many of the subtopics, including initial environmental review; identification of environmental aspects and evaluation of associated environmental impacts; internal performance criteria, objectives, and targets; and communication and reporting. Additionally, key issues to consider for major subtopics are delineated.

Avoiding Misconceptions Inherent in ISO 14004.   ISO 14004 was not given the same degree of attention in its development within TC 207 that ISO 14001 received. Consequently, it is not as consistent (internally or

Table 2-1. Comparison of Information in ISO 14001 and ISO 14004

ISO 14001		ISO 14004	
0	Introduction	0	Introduction
		0.1	Overview
		0.2	Benefits of Having an Environmental Management System
1	Scope	1	Scope
2	References	2	Normative References
3	Definitions	3	Definitions
4	Environmental Management System Requirements	4	Environmental Management System Principles and Elements
4.0	General		
4.1	Environmental Policy	4.1	How to Start: Commitment and Policy
		4.1.1	General
		4.1.2	Top Management Commitment and Leadership
		4.1.3	Initial Environmental Review
4.2	Planning	4.2	Planning
4.2.1	Environmental Aspects	4.2.1	General
4.2.2	Legal and Other Requirements	4.2.2	Identification of Environmental Aspects and Evaluation of Associated Environmental Impacts
4.2.3	Objectives and Targets	4.2.3	Legal Requirements
4.2.4	Environmental Management Programs	4.2.4	Internal Performance Criteria
		4.2.5	Environmental Objectives and Targets
		4.2.6	Environmental Management Program
4.3	Implementation and Operation	4.3	Implementation
4.3.1	Structure and Responsibility	4.3.1	General
4.3.2	Training, Awareness, and Competence	4.3.2	Ensure Capability
		4.3.2.1	Resources—Human, Physical, and Financial

Table 2-1. Comparison of Information in ISO 14001 and ISO 14004 (*Continued*)

ISO 14001		ISO 14004	
4.3.3	Communication	4.3.2.2	Environmental Management System Alignment and Integration
		4.3.2.4	Environmental Awareness and Motivation
		4.3.2.5	Knowledge, Skills, and Training
		4.3.3	Support Action
		4.3.3.1	Communication and Reporting
		4.3.3.2	Environmental Management System Documentation
		4.3.3.3	Operational Controls
4.3.4	Environmental Management System Documentation	4.3.3.4	Emergency Preparedness and Response
4.3.5	Document Control		
4.3.6	Operational Control		
4.3.7	Emergency Preparedness and Response		
4.4	Checking and Corrective Action	4.4	Measurement and Evaluation
4.4.1	Monitoring and Measurement	4.4.1	General
4.4.2	Nonconformance and Corrective and Preventive Action	4.4.2	Measuring and Monitoring (Ongoing Performance)
4.4.3	Records	4.4.3	Corrective and Preventive Action
4.4.4	Environmental Management System Audit	4.4.4	Environmental Management System Records and Information Management
		4.4.5	Audit of the Environmental Management System
4.5	Management Review	4.5	Review and Improvement
		4.5.1	General
		4.5.2	Review of the Environmental Management System
		4.5.3	Continual Improvement
	Annexes		Annexes

with ISO 14001) as it could be. Undoubtedly, the next revision will see significant modifications to this document so that it can truly be a guideline to provide assistance to organizations implementing or improving an EMS. Some obvious areas for improvement are highlighted below.

The first undesirable characteristic that permeates the document is its bias for prescriptive language, which creates the impression that ISO 14004 sets additional requirements. Repeated calls for documentation leave one wondering, "For whom is this documentation being developed?" Do managers of organizations really need to be told that they should document in order to have a good system? Is it really impossible to have a good system without documentation? Documentation is needed only when it actually helps run the system or when it is useful for registration purposes. ISO 14004 never justifies that each call for documentation is needed for better management. One is left to wonder, then, "Is this documentation needed for registration?" Well, "Why are we requiring documentation in a guideline document that is never supposed to be used for registration?"

The liberal use of the word *should* also creates the (probably) unintended impression that ISO 14004 is intent on adding more requirements to ISO 14001. The following sentence from Section 4.1.3 is an illustrative example of this prevalent problem: "The process and results of the initial environmental review should be documented and opportunities for EMS development should be identified." In this one sentence, we have demands for documentation with two uses of the word *should*, which would undoubtedly be interpreted as *must* in a third-party audit. In this case the problem is compounded, since ISO 14001 does not require initial reviews in the specification, but mentions them only in Annex A for companies with no existing EMS. Why, then, does ISO 14004 recommend something that is not required in the specification document—at least for those that already have an EMS? This ambiguity is confusing and inconsistent. To repeat, ISO 14001 determines the only requirements and is ultimately the only authority for registration purposes.

Another problem that could have been avoided is seen in the statement in the introduction, which suggests that ISO 14004 can be used for second-party recognition between contracting parties. Since ISO 14004 is not, on the whole, written as a requirements document, even this admittedly less rigorous use between two parties is problematic. How can any party claim to be in conformance with the various help sections in the document? Blurring the real purpose of this document, which is meant to assist beginners with the assurance function that's been reserved for ISO 14001, will surely baffle prospective users. Users may also be confused by unsupported statements found in the introduction, such as "an organization that has implemented an EMS can achieve significant competitive advantage." There are many good reasons to have an EMS, such as better, more consistent, and more reliable satisfaction of environmental obligations, or meeting the requirements of customers. It is not particularly useful to speculate about unsubstantiated claims like competitive advantage.

To be fair, it must be acknowledged that both EMS documents were developed in record speed compared with the usual ISO cycle. In that light, they are credible, consensus products. TC 207 will surely endeavor to improve them in their next version, which may be no later than the year 2000.

Environmental Auditing

Subcommittee 2 is responsible for environmental auditing (EA), and member bodies that make up this subcommittee are presented in Fig. 2-7. The subcommittee has drafted the following documents:

- ISO 14010, "Guidelines for Environmental Auditing—General Principles on Environmental Auditing"

- ISO 14011, "Guidelines for Environmental Auditing—Audit Procedures: Auditing of Environmental Management Systems"

- ISO 14012, "Guidelines for Environmental Auditing—Qualification Criteria for Environmental Auditors Performing Environmental Management Systems Audits"

Subcommittee 2 Members

Participating Members: Australia, Austria, Belgium, Brazil, Canada, Chile, China, Colombia, Cuba, Czech Republic, Denmark, Finland, France, Germany, Indonesia, Ireland, Italy, Jamaica, Japan, Republic of Korea, Malaysia, Mexico, Netherlands, New Zealand, Norway, Philippines, South Africa, Spain, Sweden, Switzerland, Tanzania, Thailand, Turkey, Ukraine, United Kingdom, United States, Uruguay, Zimbabwe

Correspondent Members: Argentina, Iceland, Mongolia, Poland, Portugal, Singapore, Slovakia, Sri Lanka, Trinidad and Tobago, Yugoslavia

---

Figure 2-7. TC 207 membership in Subcommittee 2 on environmental auditing, mid-1995.

These documents are to be used for guidance by registrars, auditors, and organizations implementing the ISO 14001 specification document. Elements addressed in the ISO EA documents are presented in Table 2-2.

Since many consultants were involved in the process of creating the audit documents, terminology is found within them that reflects a consultant/client orientation. However, the EMS audits called for in ISO 14001 do not require the use of a third party consultant/auditor, although use of such auditors may be deemed preferable in some cases by some organizations.

ISO 14010. ISO 14010 provides general principles on EA, and is meant to apply to all types of EA—not just EMS auditing. An environmental audit, as defined in ISO 14010, is a "systematic, documented verification process of objectively obtaining and evaluating evidence to determine whether specified environmental activities, events, conditions, management systems, or information about these matters conform with audit criteria, and communicating the results of this process to the client" (Section 3.9). The "client" is the organization that calls for the audit—usually the auditee, but sometimes another party such as corporate headquarters of a company or a government agency. This definition is much broader than the definition of EMS audit in ISO 14001. (See Chap. 8 for a discussion of EMS audit as it pertains to ISO 14001.)

ISO 14010 states that an environmental audit should have as its focus a clearly defined and documented subject matter. This concept is an important one. The auditors are not free to select what it is that they want to audit; rather, they must audit those aspects that are predetermined. This point is emphasized in several sections of the document. It is the responsibility of the client (and not the auditor) to set the audit objectives. However, the scope of the audit may be set by the auditor in consultation with the client in order to meet the client's objectives. Further, the determination of such audit criteria is an early and essential step to the process and must be agreed to by both the auditor and the client.

The document further asserts that members of the audit team should be independent of the activities that they audit. This stipulation ensures objectivity and independence of the audit process. However, use of an external or internal auditor is left to the discretion of the client.

The audit report should include numerous items, the most controversial of which is audit conclusions. An audit conclusion is defined as "professional judgment or opinion expressed by an auditor about the

Table 2-2. Elements Addressed in the ISO Environmental Auditing Documents

ISO 14010: General Principles on Environmental Auditing

0–3   Introduction, Scope, Normative References, Definitions
4   Requirements for an Environmental Audit
5   General Principles
5.1   Objectives and Scope
5.2   Objectivity, Independence, and Competence
5.3   Due Professional Care
5.4   Systematic Procedures
5.5   Audit Criteria, Evidence, and Findings
5.6   Reliability of Findings and Conclusions
5.7   Reporting

ISO 14011: Audit Procedures—Auditing of Environmental Management Systems

0–3   Introduction, Scope, Normative References, Definitions
4   Environmental Management System Audit Objectives, Roles, and
        Responsibilities
4.1   Audit Objectives
4.2   Roles, Responsibilities, and Activities
4.2.1 Lead Auditor
4.2.2 Auditor
4.4.3 Audit Team
4.2.4 Client
4.2.5 Auditee
5   Auditing
5.1   Initiating the Audit
5.1.1 Audit Scope
5.1.2 Preliminary Document Review
5.2   Preparing the Audit
5.2.1 Audit Plan
5.2.2 Audit Team Assignments
5.2.3 Working Documents
5.3   Executing the Audit
5.3.1 Opening Meeting
5.3.2 Collecting Evidence
5.3.3 Audit Findings
5.3.4 Closing Meeting with the Auditee
5.4   Audit Reports and Records
5.4.1 Audit Report Preparation
5.4.2 Report Content
5.4.3 Report Distribution
5.4.4 Record Retention
6   Audit Completion

Table 2-2. Elements Addressed in the ISO Environmental Auditing Documents (*Continued*)

ISO 14012: Qualification Criteria for Environmental Auditors Performing Environmental Management System Audits

0–3	Introduction, Scope, Normative References, Definitions
4	Education and Work Experience
5	Auditor Training
5.1	Formal Training
5.2	On-the-Job Training
6	Objective Evidence of Education, Experience, and Training
7	Personal Attributes and Skills
8	Lead Auditor
9	Maintenance of Competence
10	Due Professional Care
11	Language
	Annex A (Informative)—Evaluating the Qualifications of Environmental Auditors
A.1	General
A.2	Evaluation Process
A.3	Evaluations of Education, Work Experience, Training, and Personal Attributes
	Annex B (Informative)—Environmental Auditor Registration Body
B.1	General
B.2	Auditor Registration

subject matter of the audit, based on and limited to reasoning the auditor has applied to audit findings" (Section 3.1). Audit findings are the "results of the evaluation of the collected audit evidence compared against the agreed audit criteria" (Section 3.4). Because of the sensitive nature of environmental audits, many participants in TC 207 had reservations about allowing judgments or opinions in the audit process. They preferred that only the "audit findings," which are construed to be more objective, be put in the report. However, others felt that drawing conclusions from the findings is acceptable and in some cases desirable. This issue arises mostly during the EMS audit process and not during the registration process, which entails considerable subjectivity from the registrar as he or she tries to ascertain whether the organizations is making a sincere, good-faith effort to implement a sound EMS.

ISO 14011

Potential Confusion with ISO 14011. ISO 14011 is the guidance document used for EMS audits. EMS audits are a required element of ISO 14001, although the use of ISO 14011 is not required under ISO 14001. The optional use of ISO 14011 may cause some confusion, but it was felt that other valid audit guidelines should not be disqualified if they contain the essential elements.

Another source of confusion is the applicability of ISO 14011 to registration audits. As has already been implied, the registration audit and the EMS audit are not the same thing. They will have many similar elements, but the registrar is not bound by ISO 14011. It is expected that the registrar will use ISO 14011 during the registration audit, but other documents will also be necessary. Thus, the EMS audit as defined in both ISO 14001 and ISO 14011 and the registration audit are not synonymous.

Applicability. ISO 14011 is applicable to all types and sizes of organizations operating an EMS. Its application is similar to that of ISO 14001. Although SMEs may find ISO 14011 initially daunting, a closer look will reveal that the procedures outlined in it are reasonable and achievable.

Key Elements of ISO 14011. Key to an ISO 14011 audit is the development of an audit plan. The audit plan should be reviewed and approved by the client, and the plan should be designed to be flexible. Among other things, the audit plan should include the audit objectives and scope and the audit criteria. These suggestions are similar to those in ISO 14010.

ISO 14011 also makes it clear that the auditor is to assess the ability of the internal management review process to ensure the continuing adequacy, suitability, and effectiveness of the EMS. This again points up the fact that the auditors are to audit process and not performance. In this case "process" is the internal management review process and "performance" is the suitability and effectiveness of the system. Thus, the auditors will not be assessing the suitability and effectiveness of the system; instead, they will be assessing the internal management review process to see that it accomplishes its purpose—which is to ensure the continuing suitability and effectiveness of the EMS.

Furthermore, ISO 14011 makes it optional to include audit conclusions in the audit report. This provision allows more flexibility than that found in ISO 14010. If it is agreed that the audit report will contain audit conclusions, these conclusions will include such items as whether the system conforms to the EMS audit criteria, whether the system is properly implemented and maintained, and whether the internal review process is able to ensure the continuing suitability and effectiveness of the EMS. Once completed, the audit report is distributed as specified in the audit plan. The distribution list is determined by the client/auditee, not the auditor. Likewise, document retention is determined by agreement between the client/auditee and the lead auditor. If the auditor keeps copies of the documents, the auditor cannot disclose any document (subject to legal mandates) without permission from the client/auditee. These safeguards have been incorporated to reduce inadvertent disclosure of sensitive data that might precipitate premature enforcement actions.

Under ISO 14011, audit findings must be based on evidence and must be recorded, and all significant nonconformances must be documented. This requirement is consistent with ISO 14010. ISO 14011 adds a cautionary note that "details of findings of conformity may also be documented, but with due care to avoid any implication of absolute assurance" (Section 5.3.3). Essentially, this note is a caveat for consultants, who have concerns about liability. The fact that the organization is in conformance with its EMS does not necessarily mean that the organization is in total compliance with all laws and regulations. As already stated, an EMS audit is not expected to be a compliance audit.

When collecting evidence, the auditor is allowed to examine documents, observe activities and conditions, and conduct interviews. This last method of collecting evidence will allow the auditor to determine if employees understand their roles and responsibilities with respect to the EMS. Since employee awareness and training are important parts of the EMS, it is appropriate that auditors be given necessary access to assess whether these goals have been accomplished.

ISO 14012. ISO 14012 sets forth guidance for qualification criteria for internal and external environmental auditors who perform EMS audits. ISO 14001, the EMS specification document, does not require that ISO 14012 be used when determining qualification criteria for those auditing the EMS, but it is expected that many organizations will elect to review the guidance document and use elements from it, as appropriate to their needs. The process of certifying auditors for the purpose of performing registration audits is separate from EMS audits performed to meet requirements in ISO 14001; however, the accreditation process may also use ISO 14012 as a basis for qualifying auditors.

Education and Training Requirements. ISO 14012 specifies that auditors should have at least a secondary education, or equivalent. Those auditors with only a secondary education must also have five years of appropriate work experience. Auditors who have obtained a degree must have four years of appropriate work experience.

In addition, the auditor should have formal training in environmental science and technology, technical/environmental aspects of facility operations, requirements of environmental laws, environmental management systems and standards, and audit procedures, processes, and techniques. Formal training can be waived if other accepted methods of competence can be proved, such as through accredited examinations or relevant professional qualifications. In the view of some, the amount of formal technical training required in ISO 14012 is excessive for auditors who are to perform EMS audits. A problem can arise if technically overqualified auditors incautiously turn the EMS audit into a compliance audit.

ISO 14012 further specifies that the auditor should have completed on-the-job training for a total of 20 equivalent workdays of auditing encompassing a minimum of four audits. It requires that on-the-job training occur within a period of not more than three consecutive years. Additional requirements are specified for the lead auditor.

Personal Attributes and Skills.   Auditors are expected to possess numerous attributes and skills. These include good verbal and writing skills; good interpersonal skills such as diplomacy, tact, and the ability to listen; objectivity and independence; good organizational skills; and the ability to reach sound judgments on the basis of objective evidence.

Maintaining of Competence.   Finally, auditors are expected to maintain their competence by ensuring the currency of their knowledge through refresher courses, as needed. The auditor is responsible to maintain his or her level of experience in the execution of audits to a satisfactory degree.

A number of delegates to TC 207 expressed their unhappiness with the qualification criteria in ISO 14012 and wanted to substitute a competence requirement instead. Their argument, effectively, is that education, training, and experience do not always lead to competence and, vice versa, that competence is not always related to specific levels of education or training. It seems altogether reasonable to expect a move toward a competence measure in the first revision of this document.

Annexes.   Annex A associated with ISO 14012 provides guidance for evaluating the qualifications of environmental auditors. Annex B contains guidance on a consistent approach to the certification of environmental auditors. In effect, these annexes provide information on how to certify auditors and how to set up an auditor certification program.

Environmental Performance Evaluation

Environmental performance evaluation (EPE) is essentially a subsystem of the EMS. In ISO 14001 there is the requirement under Section 4.4.1, Monitoring and Measurement, to record information to track performance. This, in effect, calls for EPE. Once again, ISO 14001 does not specifically require that the ISO 14031 guidance document for EPE be used to meet this requirement. However, most organizations will want to review the document and use information and evaluation techniques presented there, as appropriate.

Development of the EPE Document—ISO 14031. Originally, as mentioned in Chap. 1, the Strategic Advisory Group on the Environment (SAGE), which was formed to determine if ISO should develop standards on environmental management, was interested in standardizing environmental performance. In fact, there was a U.S.-led subgroup of SAGE named Environmental Performance. It eventually became obvious during that process that the concept of standardizing environmental performance worldwide was unrealistic, and the subgroup changed its focus and its name to Environmental Performance Evaluation. Once TC 207 was formed, standardizing environmental performance was specifically excluded from the scope of work given to the committee, primarily because of the recognition that environmental performance is a public-sector issue currently under the domain of sovereign national governments. Thus, TC 207 was charged with creating a standard that addressed EPE and not environmental performance. The United States was once again given leadership responsibility for this group, which became Subcommittee 4 of TC 207. Member bodies of this subcommittee are presented in Fig. 2-8.

The original idea for the EPE subcommittee was that it would promulgate one or two guidelines to address two different aspects of performance evaluation. These aspects included generic performance indicators that would be applicable to all organizations and performance indicators for specific industries. The decision to divide up the subject matter in this way allowed both Norway and the United States to lead separate working groups under Subcommittee 4. After nearly two years of work, it was determined that this plan for EPE was not feasible, and the plan was abandoned in favor of a unified approach.

Subcommittee 4 Members

Participating Members: Austria, Australia, Belgium, Brazil, Canada, Chile, Colombia, Cuba, Czech Republic, Denmark, Finland, France, Germany, India, Indonesia, Ireland, Italy, Jamaica, Japan, Republic of Korea, Malaysia, Mexico, Netherlands, New Zealand, Norway, Philippines, South Africa, Spain, Sweden, Switzerland, Tanzania, Thailand, Turkey, Ukraine, United Kingdom, United States, Uruguay, Zimbabwe

Correspondent Members: Argentina, China, Iceland, Lithuania, Poland, Portugal, Singapore, Slovakia, Sri Lanka, Trinidad and Tobago, Yugoslavia

---

Figure 2-8. TC 207 membership in Subcommittee 4 on environmental performance evaluation, mid-1995.

The revised concept for Subcommittee 4 includes only one body of work that will address performance indicators for the EMS, performance indicators for operations in terms of effluents and efficiencies of processes, and performance indicators for the environment itself. This new approach, too, has problems associated with it. For instance, the subcommittee has had difficulty with the concept of quantifying measurements for the EMS elements. In addition, measuring both the environment and an organization's impact on the environment is problematic. Many believe that the environment is not amenable to measurement for performance evaluation, but rather is the context within which the EMS and the operational systems are evaluated. Thus, guidelines for EPE are far from complete, and concepts of what is to be included and what are proper performance indicators are still evolving.

Purpose of the EPE Guideline.   Ultimately, the EPE guideline will provide a "tool box" of environmental performance indicators. These indicators include analytical assessments that are applied to raw data. For instance, emissions can be measured and quantified, and data evaluated for trends. An indicator of performance might be reduction in emissions over time, taking into account product variability.

The guidance document on EPE will define numerous indicators that can be used by organizations of all types. The document is not designed to be comprehensive, but will serve to give examples of types of performance indicators that either can be used by themselves or can serve as a basis for developing additional, more applicable indicators. For those organizations which have not previously done EPE, it will also serve as a primer on how to approach the subject matter in a reasonable way.

The Relation of ISO 14031 to ISO 14001.   The guidance document for EPE, ISO 14031, is not an "add-on" to ISO 14001. It does not at this point and will not in the future address issues such as conducting initial reviews, defining significant impacts, and setting objectives and targets. It will not specify ways to establish an EMS. It is meant to help organizations meet the requirement under ISO 14001 for measuring results and tracking performance.

Some delegates to TC 207 would like to see the function of this document expanded for use by any institution or party that wants to evaluate the environmental performance of an organization, whether or not the organization has an EMS in place. Although ISO 14001 defines environmental performance as it relates to the EMS, the definition of environmental performance in the working draft of ISO 14031 is much broader. In ISO 14001, performance results are related to an organization's "environmental policy, objectives, and targets." ISO 14031 defines environmental performance as "the results of an organization's management of the environmental aspects of its activities, products, and services." Some argue that this definition opens the door for the document to be used more broadly, (i.e., as a way to compare one organization to another).

If the document is to have such expanded use, it would need to be much different from the one being developed to evaluate environmental performance as it relates to the EMS. Because of opposing views as to the intent of the document, work on the guideline has progressed slowly. At present, there is a lot more work to be done on this document and it is not likely to be ready for publication before 1997 or 1998.

How the Elements Pertaining to Organization Evaluation Fit Together

The EMS is the key element in organization evaluation, with environmental audit and environmental performance evaluation providing subsystem support for the EMS. This relationship is depicted in Fig. 2-9. In addition, the figure describes the key elements of the EMS and, for multifacility enterprises, shows a dotted-line dependence on corporate management.

Product Evaluation

The ISO 14000 series includes a number of tools for specialized uses. The guide for environmental aspects in product standards (EAPS) and the standards for environmental labeling and life cycle assessment are tools to be used by practitioners and specialists in those fields. They are not necessarily part of an EMS, and are not required under ISO 14001, although labeling and life cycle assessment guidelines will surely be useful to organizational EMS managers.

There is some confusion that product evaluation is required in order to become registered to the ISO 14001 standard. Product evaluation is not required in order to become registered. These product evaluation tools may provide some basic guidance that can be used internally by an organization, but for the most part they are meant for a select group of experts who are performing product evaluation functions today.

Guide for Environmental Aspects in Product Standards (EAPS)

ISO 14060, recently renumbered Guide 64 and entitled "Environmental Aspects in Product Standards," is an ISO guide to be used by all ISO

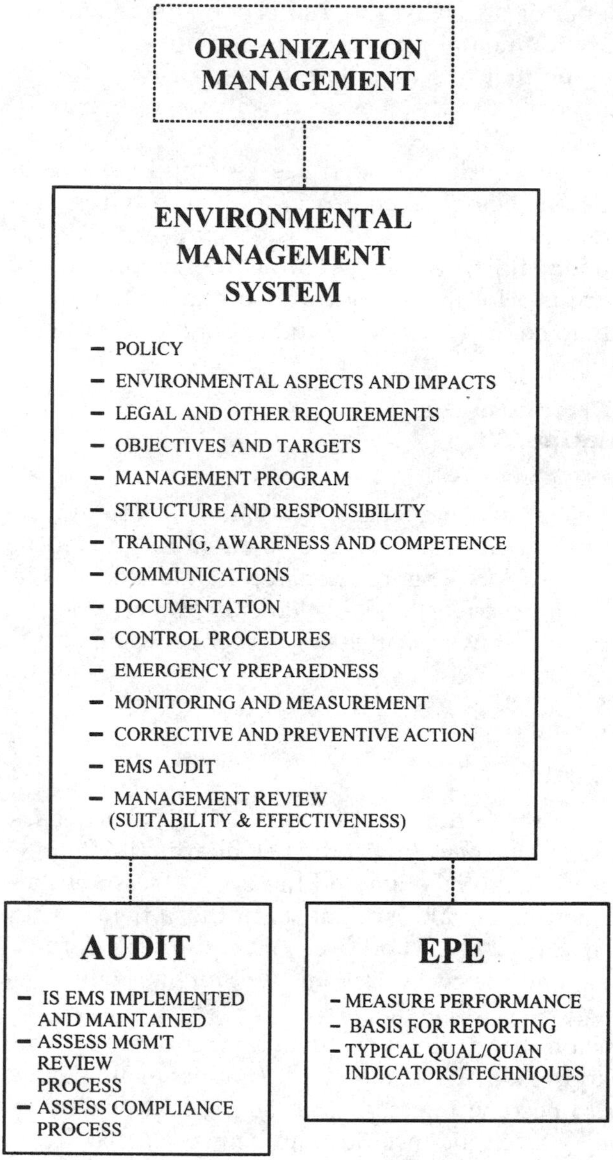

Figure 2-9. Relationship between the EMS and its sub-systems.

standards writers. SAGE recognized early in the ISO 14000 process that it might be beneficial for standards writers to have an environmental guide when writing product standards. This guide would help standards writers avoid inserting specifications in product standards that could turn out to be environmentally detrimental. In addition, the guide was to suggest characteristics that standards writers could incorporate into standards that would improve the environmental profile of a product.

The creation of the guide turned out to be more difficult than expected. At the onset, several approaches to writing the guide were tried and subsequently abandoned because of technical difficulty or other reasons. For instance, early on, the creators of the guide thought that all chemicals and substances that are deemed harmful to the environment should be delineated in the document. This approach quickly became problematic. Questions about safe-threshold limits, control methods, beneficial uses, and other scientific considerations made a simple listing impossible.

Another approach that was considered was to make the guide a design for environment (DFE) primer, specifying DFE principles for standards writers. This approach was also dropped because of the complexity of the subject matter. There are literally hundreds of industries which could be addressed in such a document, as well as numerous elements involved with the topic of DFE, including design for manufacture, design for recycle, design for end of life, and design for disassembly. Some technical organizations and agencies, such as the International Electrotechnical Commission (IEC) and the U.S. Environmental Protection Agency (EPA), have embarked on the creation of DFE guides for specific fields. In the case of the IEC, Guide 109 for the electronics industry addresses the incorporation of environmental aspects in electronics products standards. EPA has published a series of pollution prevention design documents for numerous industries that have elements such as source reduction during manufacture, recyclability, and other DFE concepts. These guides have had some success, since their scope is relatively narrow. A generic DFE guide is not practical, however, for the potentially huge array of ISO products.

Some drafters of the EAPS guide wanted product standards writers to perform a life cycle assessment (LCA) before writing a standard for a product. The thinking here was that through LCA, the standards writers would have the benefit of knowing which materials, which elements, and which characteristics to include or not include in the product so that it would be more environmentally benign. Other delegates to TC 207 thought it unreasonable to expect product experts to undertake a complex LCA prior to writing a standard. Further, LCA was not considered a mature technique that could be recommended for general-purpose use.

Ultimately, the approach adopted for the EAPS guide rests on warning standards writers that the specifications put into product standards will have environmental consequences, whether positive or negative, and that they should consider these carefully when developing criteria, elements, and characteristics that go into the standards. In addition, the guide asks standards writers to seek the assistance of experts in LCA, DFE, and other environmental areas when writing a standard. Thus, the guide takes a modest approach, limiting its scope and content, but providing a reasonable approach to a very complex, if not altogether controversial, subject.

Environmental Labeling

Environmental labeling (EL) programs have emerged over the last 15 years, and there are now approximately two dozen national programs worldwide. In 1979, Germany was the first to initiate this type of program, called the Blue Angel program. During the 1980s numerous other countries, including the United States, the Netherlands, and Canada, established EL programs. The main issue with these programs is that they are idiosyncratic and inconsistent with one another. Each has its own approach to the development of environmental criteria and each has its own goals and objectives for its program. Major corporations have found that these programs are confusing, inconsistent, and nonscientific. As a result, many corporations are hesitant to make use of them. For organizations that do business in many countries, compliance to multiple labeling requirements is also complicated, costly, and administratively burdensome.

It was therefore appropriate for ISO to include EL standards in its series of environmental management standards. An international labeling standard can have a significant positive impact on environmental labeling. It can provide a consistent approach and uniform rigor in the development of criteria for labeling. With an international standard, labeling programs have a chance to obtain needed respectability, and, hopefully, once the standard is implemented worldwide, industry will be more willing to participate in labeling programs.

Subcommittee 3 members, which are presented in Fig. 2-10, are developing five documents. One document is ISO 14020, entitled "Principles of All Environmental Labeling." This document is meant to provide guidance on the goals and principles that should be consistently incorporated into all types of environmental labeling programs. Another labeling document, ISO 14021, is entitled "Environmental

---

Subcommittee 3 Members

Participating Members: Australia, Austria, Belgium, Brazil, Canada, Chile, China, Colombia, Czech Republic, Denmark, Finland, France, Germany, India, Indonesia, Ireland, Japan, Republic of Korea, Malaysia, Mexico, Netherlands, New Zealand, Norway, Philippines, Russian Federation, Singapore, South Africa, Spain, Sweden, Switzerland, Tanzania, Thailand, Trinidad and Tobago, Turkey, Ukraine, United Kingdom, United States, Uruguay, Zimbabwe

Correspondent Members: Argentina, Cuba, Estonia, Greece, Iceland, Indonesia, Poland, Sri Lanka, Yugoslavia

---

Figure 2-10. TC 207 membership in Subcommittee 3 on environmental labeling, mid-1995.

Labeling—Self-Declaration of Environmental Claims: Terms and Definitions." This document is meant to apply to manufacturers who are declaring that their product has an environmental attribute—for example, that it is recyclable, energy-efficient, or made without ozone-depleting chemicals.

The third document is ISO 14022, entitled "Environmental Labeling—Symbols," and the fourth, ISO 14023, is entitled "Environmental Labeling—Testing and Verification Methodologies. The fifth document under development is ISO 14024, entitled "Environmental Labeling—Practitioner Programs: Guiding Principles, Practices and Certification Procedures for Multiple Criteria (Type I) Programs." This document applies to traditional third-party labeling programs, or "seal" programs.

ISO 14020. A key issue originally associated with the subcommittee work on ISO 14020 was the inclusion of a goal statement in the introduction. This proved to be very controversial and was subsequently deleted. One position was that EL should have as its goal the improvement of environmental attributes of products. This line of thinking developed because some programs, such as the seal programs, have product improvement as a goal, and they specifically use EL as part of their program. However, this does not necessarily mean that the universal goal of EL, per se, should be to improve the environmental attributes of products. In fact, it can be argued that a manufacturer may use EL for the sole purpose of competing in the marketplace, and not for the purpose of product improvement.

A second position, and one that the authors share, is that EL should be used to inform the public about relevant product characteristics. The public can use the information to make informed choices about whether to buy that product. An indirect result of increased communication about product characteristics may be improvement of the environmental attributes of products. Such product improvement would be a desirable benefit, but not the goal of EL.

ISO 14021.   Manufacturer labels tend to be single-claim labels—such as the claim that a product uses recycled materials—although there is no prohibition to displaying multiple claims. The objective of creating an ISO standard for this type of labeling is to ensure that the information on the label is accurate, verifiable, and nondeceptive. This last characteristic is important, since information that is accurate and verifiable can still be deceptive.

Deception occurs when information about an environmentally beneficial attribute is highlighted, but is actually of lesser significance than another product attribute that is environmentally problematic. Thus, making a single-attribute claim about a product, such as "made with 100 percent recycled materials," is inherently deceptive, even if true, if the product also contains toxic materials. In order to ensure that single-attribute claims are not deceptive, the claims on the label must be nontrivial when considered in the context of the entire product. Exact requirements in the standard to effectuate this outcome are currently being developed. They include discussion of what consideration to accord LCA as part of the requirements for manufacturer claims. At this point it seems impractical, from both cost and technical perspectives, to ask manufacturers to perform LCAs on their products for the purpose of making single-attribute claims.

ISO 14021 is still burdened with the same product improvement goal that has been deleted from ISO 14020. The process of educating newly active delegates to TC 207 is ongoing as of year-end 1995.

ISO 14024.   Third-party labeling programs, or seal programs, use criteria based on many product characteristics and attributes to determine, in effect, which products have overall environmental superiority. A seal is awarded to those products that meet the specified environmental criteria so that consumers will know which products are the most environmentally favorable. The purpose of ISO 14024 is to lay out principles and protocols that labeling programs can follow when developing environmental criteria for a particular product. This uniform approach will lead to greater agreement among stakeholders and less divergence in product criteria developed by different programs. The desire and expectation is that the product criteria developed by different programs will be so similar that it will no longer make a difference where the criteria are developed.

Product criteria that are used for comparing one product to another must support a claim of overall superiority. The criteria must be rigorous, because the organization that is awarding the seal is telling the public that the product with the seal is environmentally superior to its competitor. In order for an organization to make this type of determination, the approval process should be based on formalized analysis and scientific certainty that the product selected for the seal is truly superior in terms of its environmental characteristics.

At this point, the expert delegates feel that product LCA or some similarly rigorous discipline must be part of the evaluation process for seal programs, regardless of the cost and the technical burden. There is a general consensus that because of the trade ramifications associated with seal programs, the evaluation must be performed in a very scientific way, which must include LCA. On the other hand, there is a feeling that very complex products—such as computers, refrigerators, televisions, and automobiles—are not amenable to LCA, and therefore these types of products should not be included in EL programs. Even simple product comparisons, such as disposable versus cloth diapers, are still under debate in terms of their life cycle impact on the environment.

Once the principles and requirements of seal programs are developed, it is desirable for TC 207 to make ISO 14024 a specification standard. This would mean that organizations that create EL programs would have the opportunity to obtain certification to ISO 14024. Becoming certified to ISO 14024 would greatly benefit a seal program, since, once certified, the program would gain international respectability. Its output would be deemed reliable and acceptable by industry and other affected parties. Ultimately, once certified programs prove consistency among themselves, there will be greater consumer confidence in EL, and industry will seek such labeling for their products. This should lead to environmentally conscious consumption and overall environmental improvement.

International Trade.   EL has the potential to create serious international trade issues through its power to discriminate against products on the basis of environmental preferability. The recently established Technical Barriers to Trade (TBT) Agreement, under the World Trade Organization (WTO), does allow technical barriers to trade in certain situations. The barriers must be created in a nondiscriminatory manner and in an open, consensus process that allows all stakeholders and interested parties to participate.

Before the TBT Agreement was enacted, discrimination against products had been allowed solely on characteristics of the product and on what the product contained, but not on how it was manufactured or where it was manufactured. This was based on the requirements outlined in the Production and Process Methods (PPM) Rule. Under the TBT Agreement, however, discrimination can also rest on how or where a product is manufactured, as long as a consensus, nondiscriminatory process is followed.

The TBT Agreement directly applies to and affects seal programs. If a third-party seal program has been developed in an open, consensus process that conforms to the TBT Agreement, then the use of the seal is allowed, even though it may lead to discrimination against some products. Also, since LCA may be used as part of the scientific evaluation of a product, and since it takes into account how and where the product was manufactured, there may be discrimination based on PPM factors as well. This too will be deemed acceptable, as long as the rules of the TBT Agreement have been adhered to.

Further, it follows that if ISO 14024—which sets forth criteria for the seal programs—prescribes a process that meets the requirements of the TBT Agreement, then any seal program that meets the requirements of ISO 14024 will be valid from an international trade perspective, even if the program discriminates against some products. Thus, the ISO subcommittee must ensure that all requirements under the TBT Agreement are incorporated into the standard in a straightforward and rigorous way, in order to avoid inappropriate discrimination through use of a lax standard. In addition, the standard, through its principles, protocols, and procedures, must ensure that the criteria used for seal programs are scientifically based and consistent. In this way, seal programs that adhere to the ISO 14024 standard can gain international respectability and validity. This is a challenging task for the delegates to the ISO subcommittee, and it may be necessary to bring in trade experts and other technical professionals to accomplish the desired objectives for seal and other EL applications.

LCA

Subcommittee 5 of TC 207 is responsible for the development of standards on LCA. Members of this subcommittee are presented in Fig. 2-11.

Originally, LCA standards were termed *life cycle analysis*. The name was changed recently because life cycle analysis implies a rigorous, scientific process, used to evaluate the environmental impact of all aspects of a product—its materials, method of manufacture, use, disposal, and other applicable elements. *Life cycle assessment* is a less rigorous and less scientific process, and since the state of the art is still

---

Subcommittee 5 Members

Participating Members: Argentina, Australia, Austria, Belgium, Brazil, Canada, Chile, Colombia, Denmark, Finland, France, Germany, Indonesia, Israel, Italy, Japan, Republic of Korea, Malaysia, Mexico, Mongolia, Netherlands, New Zealand, Norway, Philippines, South Africa, Spain, Sweden, Switzerland, Thailand, Turkey, Ukraine, United Kingdom, United States, Uruguay, Zimbabwe

Correspondent Members: China, Cuba, Czech Republic, Iceland, Ireland, New Zealand, Poland, Singapore, Slovakia, Sri Lanka, Trinidad and Tobago, Yugoslavia

---

Figure 2-11. TC 207 membership in Subcommittee 5 on life cycle assessment, mid-1995.

not at a scientific level, this term seems to more appropriately reflect current reality.

Draft LCA Standards.   Several standards are being drafted to provide guidance on the confusing, often controversial, topic of LCA. ISO 14040 is entitled "Life Cycle Assessment—Principles and Framework." The purpose of this document is to provide a clear overview of the practice, applications, and limitations of LCA to a broad range of potential LCA users and stakeholders, some of whom may have limited knowledge of LCA. The list of limitations in this document is so daunting that one is left to wonder what practical use LCAs can ever really have.

ISO 14041 is entitled "Life Cycle Assessment—Goals and Definition/ Scope and Inventory Analysis," and is intended to describe special requirements and guidelines for the preparation, conduct, and critical review of the life cycle inventory analysis. Inventory analysis is the phase of LCA that involves the compilation and quantification of environmentally relevant inputs and outputs of a product system. ISO 14042, entitled "Life Cycle Assessment—Impact Assessment," proposes to provide guidance on the impact assessment phase of LCA. This phase of LCA is aimed at evaluating the significance of potential environmental impacts using the results of the life cycle inventory analysis. However, the methodological and scientific framework for impact assessment is still being developed. Because of the inherent subjectivity in impact assessments, the most critical requirement for their conduct will be transparency, so that decisions and assumptions can be clearly described and reported.

The last document in the series, ISO 14043, is entitled "Life Cycle Assessment—Improvement Assessment," with a given scope to improve the total environmental performance of product systems. It has been suggested that this last document should rather address "interpretations" of LCA results in relation to the goal definition phase of the study. This may involve reviewing the scope of the LCA, as well as the nature and quality of the data collected. Assuming that the existing problems can be solved, LCA may be a useful supporting tool for EMS—in particular, for support of other activities covered by standards in the ISO 14000 series. These include standards for EPE (ISO 14031), EL (ISO 14020, 14021, and 14024), and the EAPS guide (ISO Guide 64).

Standardization of LCAs. The challenges facing the standardization of the LCA process are similar to the challenges found in standardizing criteria for environmental labeling. Currently, product LCAs are being performed in many countries around the world—in Norway, Sweden, Denmark, France, the United Kingdom, Japan, Canada, and the United States, among others—and there is no uniform approach or methodology criteria for performing these assessments.

Since there is no accepted scientific approach to LCA, every practitioner approaches it in his or her own manner. The choice of calculation methods, the quality of the data used, the system boundaries that are defined for the problem, and the assumptions that are made during the study are all approached differently depending on which practitioner is performing the LCA. The result is inconsistency, unpredictability, and unreliability in the practice of LCA. Further, there is a possibility (if not a reality) that the assumptions made are, at best, nonscientific and, at worst, prejudicial in nature. Use of such assumptions can lead to discrimination against certain products, and can result in unfair trade practices at the international level. If misused, the LCA process can prejudicially disparage a particular product or, conversely, can set up a competitive advantage for a product in the marketplace.

The subcommittee drafting the LCA standards is trying to provide requirements that will result in a consistent, predictable, reliable, and scientifically based approach to LCA. The goal is to minimize erroneous, prejudicial conclusions and inconsistent results in assessments. Most of the delegates on the subcommittee consider the challenge to be monumental, and some consider it to be almost impossible.

In particular, there has been difficulty in developing a scientific methodology that addresses environmental impacts, mainly because of the complexity of the subject matter. Certainly, the subject matter is too complex for simple models, and detailed models that accurately and scientifically address environmental impacts are nonexistent. In fact, the issue of what constitutes an environmental impact has been debated by scientific experts during the last quarter century, and there have been very few agreements about what constitutes significant impact and/or harm to the environment.

Practitioners of LCAs have had little guidance by which to formulate a sound methodology or approach to their assessments. Thus, the development of criteria, procedures, and protocols for LCA, which can make the LCA practice more consistent and reliable, are an important part of ISO 14000, and will continue to be a focus of TC 207. Once rigorous, scientifically based methodologies for the LCA process are developed, the practice will gain in respectability and validity.

Conceptual Approach to LCAs.   Some delegates to Subcommittee 5 feel that a conceptual approach to LCA is adequate and that the use of this approach should be recommended in the ISO standard. Although most experts agree that a conceptual approach is too simplistic for use in third-party LCAs, it does have some applicability for internal LCAs.

For instance, an organization might use a conceptual LCA when selecting materials. The materials can be viewed comparatively as to their environmental consequences, with a conceptual review of their recyclability potential, toxicity, and other factors deemed important by the organization. In addition, conceptual LCAs can be used in making policy decisions. This approach can, in a general way, help an organization test one option against another and can safely be used for establishing policies and goals.

A conceptual approach, however, is not rigorous or scientific enough to be used when comparing one product against another for the purpose of labeling, advertising, or otherwise providing a market advantage. Since fair trade practices are at stake, LCAs for this purpose must be consistent, scientific, and reliable.

Other Concerns Associated with the Use of LCAs.   There are some specific concerns associated with the standardization and use of LCAs that should be addressed by Subcommittee 5. One concern is that, at this point, the state-of-the-art process for LCAs is not sophisticated enough to be useful to evaluate complex products such as computers, televisions, telephones, and automobiles. This limitation needs to be acknowledged somehow within the standard.

In addition, if a comparison of product function is to be part of the LCA, it must be carefully handled in the standard. Some products may appear to perform the same function, but in reality do not. This could be a source of confusion and bias when performing LCAs. Further, there are concerns about the use of data in the process. Are the data comparable in substance and quality from one LCA to another? How to ensure comparability of the data needs to be addressed in the standard.

Another concern brought forth by some TC 207 delegates pertains to the role of improvement assessment in the full LCA process. Some parties insist that the goal of the LCA is to improve the product. Thus, they feel that improvement assessment is an intrinsic part of the LCA. There are many others, however, who insist that the LCA is independent of actions taken to improve the product. In addition, some contend that the LCA science is too new to be able to define improvement methodologies. This is the reason for the suggestion to change this phase to an "interpretation" phase. This debate will have to be settled within Subcommittee 5.

Finally, there is the concern that the creation of an LCA standard that performs the function of ensuring a consistent, uniform, scientific, unbiased LCA process is unachievable. If this is the case, it would be preferable for the subcommittee to abandon the project of an international LCA standard rather than develop something that is weak and ineffectual, and that may serve to validate inconsistent approaches or unacceptable practices currently used in LCA studies. Having a poor standard is clearly less desirable in this case than having no standard at all.

3

Why Is ISO 14000
Important?

*Our companies, aided by strong government
support, are looking at ISO 14001 as an
important agent to promote competitiveness
in national and international markets.*

MAURICIO REIS
*Sustainable Development General Manager,
Companhia Vale Do Rio Doce, and Leader,
Brazilian Delegation to ISO TC 207*

It is expected that thousands of organizations worldwide will spend
time and money to implement ISO 14000 in the next several years. And
tens of thousands of organizations will become registered to the standard over the next decade. Skeptics may ask, "Why spend time and
money on voluntary standards?" or "What's in it for me?" Reasons for
organizations to implement ISO 14000 are numerous and varied, and
this chapter discusses the most important ones.

The ISO 14000 standards will be a factor in international development and commerce for numerous reasons, three of which are key.
First, the standards facilitate trade and remove trade barriers; second,
the creation of the standards will improve environmental performance
worldwide; and third, these standards build worldwide consensus that
there is a need for environmental management and for a common terminology for environmental management systems. This chapter focuses on these key reasons, and also discusses other ways that ISO 14000
will influence commerce and improve environmental care globally.

Trade

Trade Barriers

From the very beginning, the ISO 14000 standards gained strong support from industry, because they promised to facilitate trade and remove trade barriers. In recent years, there has been a proliferation of national and regional standards in the environmental field. Examples include the eco-labeling programs introduced in some two dozen countries during the last decade; various European management standards similar to British Standard (BS) 7750; a veritable flood of standards from the Canadian Standards Association (CSA) for environmental management, auditing, labeling, design for environment, risk assessment, and procurement; and the European Union's (EU's) Eco-Label Regulation and Eco-Management and Audit Scheme Regulation (EMAS)—both of which rely heavily on consensus standards for their operational parameters. The United States, too, has published dozens of technical standards under the sponsorship of the American Society of Testing and Materials (ASTM) to address needs in environmental testing and monitoring associated with emission and effluent controls. This proliferation of national and regional standards has led to confusion, at best, and to trade barriers, at worst.

An example of how national and regional standards can create trade barriers can be seen in the EU's Eco-Label Regulation. That directive seeks to encourage consumer preference for environmentally sound products by awarding product labels to those from their class that satisfy a set of environmental criteria. Those products that receive the label are expected to meet basic "minimum" requirements, as delineated in Fig. 3-1.

European Union's Eco-Label Regulation

✓ Products must conform to the European Union's health, safety, and environmental legislation, wherever manufacture takes place.

✓ Products must not contain substances injurious to health or the environment, whether the substances are inert or not.

✓ Manufacturing processes must conform to European Union's standards, wherever manufacture takes place.

✓ The product must be made by state-of-the-art manufacturing processes.

Figure 3-1. Minimum requirements for a European union eco-label.

If applied strictly, these requirements have the potential to bar most manufacturers from developing countries. It will be extremely difficult for many such manufacturers to conform to EU standards and to employ state-of-the-art processes in their countries. The cost alone, assuming qualifying technologies are available to them, will prove prohibitive. Auditing of companies outside the EU for conformance to the specified criteria will be an additional burden and may itself lead to product exclusions. Also, since both product criteria and label awards are determined in the EU, the possibility of national bias in favor of local industry cannot be discounted.

As international consensus standards, the ISO 14000 documents will serve to unify countries in their approach to eco-labeling, environmental management, and life cycle assessment. A unified approach will remove trade barriers so as to facilitate trade. In addition, the drafters of ISO 14000 have been very careful in the wording and intent of the standards to ensure that they themselves will not unduly or unnecessarily hamper trade.

Trade Agreements and Trade Sanctions

ISO 14000 has the potential to play a major role in shaping the application of environmental considerations in international trade agreements. Already, environmental considerations have had increasing play in recent international trade negotiations. The Uruguay Round on General Agreement on Tariffs and Trade (GATT), the newly approved World Trade Organization (WTO), and the North American Free Trade Agreement (NAFTA) have all had extended negotiations on the link between the environment and trade. Consequently, the search for ideas and approaches to the resolution of the natural tension between these two goals has emerged as a significant activity among trade and economics experts.

Briefly, the debate at the international level considers the acceptability of using trade sanctions against parties that do not conform to the environmental expectations or standards of other trading parties or of the global community at large. Countries with the higher standards will often strive to protect either their own environmental quality, that of the global commons, or in some cases that of the "less concerned" trading partner itself. Negotiations have proved very difficult so far, for many reasons, including ones that touch on questions of national sovereignty, the lack of scientific consensus, and the general reluctance to tie enormous advantages of free trade to what some parties believe are political or subjective views on quality-of-life issues, cultural values, or (some claim) the heightened sensitivities of environmental activists.*

Often, the standards advanced during these discussions deal with environmental performance levels. Basically, some parties ask other parties to meet some environmental performance level. The attainment of such performance levels might call for the use of a specific technology, specific product design, changes in production methods, or specific requirements for process emissions. Demands for these types of achievements in trade negotiations make discussions very complex and, usually, end in stalemates.

ISO 14000 offers a more promising approach to the resolution of issues involving trade and environment. Specifically, because of what it requires, ISO 14001 can be used as an indicator of a country's desire and commitment to foster environmental protection through better environmental management in its organizations and enterprises. The advantage of this approach is compelling, for it avoids all the above-mentioned pitfalls of setting and imposing externally established performance levels. The usual objections of sovereignty, scientific consensus, and cultural values disappear, since these are not in play with management systems—certainly not nearly to the degree they are with performance standards.

The prediction is that successful promotion of ISO 14001 within countries will lead to the kind of environmental progress that can be reassuring to the world community. The standard requires organizations to know, understand, and make good-faith efforts to comply with their laws and regulations. It asks for the allocation of resources, personnel, and management focus to systematize environmental care. It promotes audits, performance measurement, and management review, as well as third-party assessments that provide assurance that these requirements are being planned and executed in good faith and with success. All of this can be accomplished without imposing performance edicts from outside and without touching on sensitive matters of national pride or sovereignty. It seems reasonable to the authors that, as ISO 14001 is implemented worldwide, registration to the standard will increasingly satisfy requirements for environmental protection in trade discussions and agreements.

---

*The claim about heightened sensitivities of environmental activists is aimed at the United States and stems largely from the U.S. position on saving dolphins—a position not universally shared.

Achieving Consensus on a New Environmental Ethic

Another reason the ISO 14000 standards are important is that they promote the practice of environmental management on a worldwide basis. They also lead to a much improved level of understanding and ability to communicate internationally on environmental management and care.

The Need for International Standards

Prior to the ISO 14000 standards, there were few environmental management standards, and certainly none that were recognized as consensus standards by all countries. The mere fact that international consensus has been attained on this relatively sensitive issue is both remarkable and consequential in more ways that the drafters might have ever imagined. As such, ISO 14000 standards will play a significant role in the environmental evolution of the planet. Fortuitously, they arrived at a particularly propitious time in light of the international trade needs mentioned earlier, as well as the growing realization that command-and-control regulation may have run its course and that further progress calls for a fundamental change in strategy.

Common Terminology

In addition to filling the need for international consensus standards, ISO 14000 provides us with a common environmental terminology. Arguably, ISO 14000 has established the "lingua franca" of environmental management. As the standards are adopted internationally, it will be much easier to go to any country in the world and discuss environmental management and any of the elements within the ISO 14000 standards, such as management review, environmental performance evaluation, audits, and other environmental systems and tools. These concepts are being defined at the international level, enabling people of all countries to speak to each other about environmental management, to share ideas for improvements, and to get workers to focus on the environmental aspects of their work. Achieving uniform understanding of environmental management terms and concepts will make international harmonization of environmental strategies a possibility. It will also promote more uniform environmental progress in all countries.

ISO 14001: Better Environmental Management (at an Affordable Price!)

ISO 14001 provides organizations with a framework for achieving more consistent and reliable environmental management. The ISO 14001 specification outlines an environmental management system (EMS) designed to address all facets of an organization's operations, products, and services. Some of these elements include environmental policy, resources, training, operations, emergency response, audits, measurements, and management reviews. The system's approach recognizes that the manner in which an organization protects the environment is as important as the goals it is expected to meet. In fact, how organizations go about meeting those requirements determines whether or not they can consistently succeed in protecting the environment and complying with existing regulations.

Leading industrial enterprises have learned, over time, that only by systematizing and integrating environmental protection into overall management can they achieve affordable, consistent compliance with internal and external requirements. Unfortunately, this lesson has often been learned, at great expense, after experiencing a serious environmental incident. The problem typically stemmed from a failure to systematize and fully integrate environmental protection into all operations—either because of institutional inertia or because of entrenched beliefs that compliance can be achieved by addressing specific points of vulnerability, installing technological controls, or assigning regulatory and compliance issues to internal specialists. It was believed that by installing sufficient controls and using professionals, an organization would achieve excellent levels of regulatory compliance. In fact, this approach to environmental compliance has worked well for some organizations. In the long run, however, it can be very expensive and wasteful. In addition, it is often unreliable and, in many cases, unaffordable.

The requirement in ISO 14001 to build and operate an EMS focuses the organization's efforts on establishing reliable, affordable, and consistent approaches to environmental protection that engage all employees in the enterprise. The environmental protection system becomes part of the total management system, receiving the same attention as quality, personnel, cost control, maintenance, and production functions. Reliability is achieved through continual awareness and competence of all employees, rather than through extraordinary or isolated efforts of specialists. Thus, ISO 14001 has the potential to provide consistent environmental protection through better management, at an affordable price.

Cultural Changes Within the Organization

The implementation of environmental management systems will bring gradual cultural change within organizations and, hopefully, throughout the world. This is a reasonable expectation, since the standard requires increased awareness, education, training, and care from employees so that they understand and respond to the environmental consequences of their work. In addition, each employee is required to adhere to the environmental policy of the organization and to know how he or she can avoid or minimize environmental incidents. The involvement of all employees in the environmental management process promotes an environmentally conscious culture in the organization and, hopefully, in the private lives of individuals too.

By implementing ISO 14001, an organization becomes aware of its environmental aspects. Environmental aspects include all elements of an organization's activities, products, and services that can have an impact on or interact with the environment. ISO 14001 asks organizations to take note of all their environmental aspects to determine which ones have or may have significant environmental impacts.

While it is true that heavily regulated organizations may have already identified their environmental aspects to satisfy applicable legal requirements, this is certainly not the case for all organizations in all countries. The failure of many governments to enforce their own national regulations means that few organizations worldwide even undertake this kind of self-analysis. The requirement in ISO 14001 to catalog significant environmental aspects redresses this deficiency and leads to a heightened awareness of the environmental dimension of operations, products, and services within organizations. Awareness is, of course, the first step and a necessary precursor to responsible operations and growth toward environmental stewardship. Once organizations begin thinking about all their environmental aspects, and not just those aspects that are regulated, the organization's approach to environmental protection will take on a whole new dimension.

To a large extent, organizations in the United States initiated the process of understanding their environmental aspects through the toxic release inventory (TRI) reports required under Title III of the Superfund Amendments and Reauthorization Act (SARA). TRI reports catalog releases to the environment of over 300 chemicals. Before the reports were required, organizations or facilities had limited knowledge of their nonregulated emissions to air, land, and water. The TRI experience will make it easier for U.S. firms to investigate their environmental aspects for ISO 14001.

Other elements required in the ISO 14001 specification provide additional impetus to improve an organization's environmental protection culture. For instance, environmental auditing requires employees to pool their efforts and their knowledge in order to satisfy audit requirements. These concerted efforts will increase awareness, coordination, cooperation, and mutually supporting actions. Additionally, the required monitoring and measurement of environmental performance, as well as the management review of data and results, provides a constant reminder to employees that management is interested in the continual effectiveness and improvement of the EMS.

The requirement to maintain knowledge of applicable laws and regulations and to keep compliance processes viable and effective involves effort and focus from all members of the organization, not just the environmental professionals. In short, the requirements of ISO 14001 will work to promote a more environmentally enlightened, aware, and sensitive organizational culture. The importance of heightened awareness for undergirding and improving the environmental performance of any enterprise cannot be overstated.

Environmental Performance

As a result of improved environmental management practices, as well as heightened employee awareness and sensitivity to environmental care, the authors contend that ISO 14001 will improve environmental performance worldwide. Skeptics may question this belief, since the focus of the standard is the implementation and continual improvement of an organization's EMS and not environmental performance. However, drafters of the ISO 14001 standards believe that focused management on an organization's environmental aspects will result in better environmental performance.

The ISO 9000 quality standards have proved this concept in practice. Organizations that have implemented the ISO 9000 standards will attest that the quality of the end product improved even though product quality was not specifically addressed in the standard. Although product improvement may not be dramatic, achieving consistency of quality is itself a quality improvement.

Of course, the EMS standard is not designed to achieve any particular level of performance (e.g., technology or effluent levels). But, through its use, an organization can be assured that its ability to meet its environmental obligations is maintained, and that incidents or excursions are limited or avoided altogether. The result is that overall environmental performance will, in fact, be improved. Better performance and reliability result in reduced liabilities, more satisfied constituencies, and an improved environment.

Other Reasons ISO 14001 Is Important

Effect on the Banking Industry

The application of ISO 14001 to qualify prospective recipients of bank loans and aid for development projects has not yet been explored. International financial institutions—such as the World Bank, the International Monetary Fund, the Export-Import Bank, the Overseas Private Investment Corporation, the United States Agency for International Development—as well as private-sector commercial lenders and equity investors may eventually require ISO 14001 commitments from borrowers. Since ISO 14001 can be a credible indicator of an organization's efforts to meet its environmental responsibilities, it seems reasonable that registration to the standard could be used to "screen" prospective borrowers and recipients.

Awareness of Applicable Laws and Regulations

ISO 14001 requires an organization to be aware of all environmental laws and regulations applicable to its environmental aspects. This requirement will compensate, to a considerable extent, for the ignorance that prevails in places where such laws are not enforced. Today, many organizations throughout the world have only a vague notion of the laws they are subject to. ISO 14001 may also lead some countries to discover that they have many more laws on their books than they can ever enforce, given their resources. Whereas in past years developing countries were encouraged to adopt environmental laws from more-developed countries, compliance and enforcement may have become challenges that strain both the societal commitment and the institutional capacity for proper execution.

In other instances, a country may have appropriate regulations but not the infrastructure for effective implementation. This is a structural problem that cannot be addressed through a management standard alone. However, awareness of applicable laws is the first step in the right direction, and it may, through its own compelling dynamic, spur evolutionary changes in behavior, technological investment, and institutional will to build the necessary infrastructure.

Promotion of Processes to
Maintain Regulatory Compliance

ISO 14001 is expected to promote the development of processes to maintain environmental compliance. While compliance with all applicable laws may be difficult or elusive in many countries, ISO 14001 expects organizations to implement processes to maintain such compliance. In countries where enforcement is strict, compliance processes are a part of doing business and can simply be integrated into the overall management system. In countries where enforcement is either lacking or ineffectual, ISO 14001 will provide the needed (and in some cases the only) impetus to develop processes to reach and maintain compliance. In effect, the standard encourages compliance processes, even in countries where compliance and enforcement have not traditionally been strong. Of course, knowledge of the applicable laws is a prerequisite for establishing any compliance process.

In some developing countries, compliance options will be limited by deficiencies in both organizational resources and available infrastructure. As noted above, infrastructure plays a key role in compliance, since it is very difficult to be in regulatory compliance without the necessary infrastructure. For instance, if there are no recycling facilities in an area, a law that requires recycling is difficult, if not impossible, to comply with. In these cases, organizations may be disadvantaged in meeting the requirements of ISO 14001, since their implementation of credible compliance processes may require greater efforts to overcome structural national deficiencies. If there are no reasonable ways to be in compliance with specific country laws, an organization will not be able to implement a compliance process to meet those laws.

Conceivably, this situation may provide impetus for some countries to redraft their environmental laws so that they match their existing resources and capabilities. Although redrafting laws to match resources and capabilities may weaken the legal framework in the short term, the overall effect is to increase the ability of organizations to comply with legal requirements. As the infrastructure of a country improves, laws can be made progressively strict. The overall effect is to increase the credibility of all parties involved with environmental progress, including legislators, organizations, and enforcement authorities.

On the other hand, countries with an economy that is strong enough to provide an environmental infrastructure should opt to build this infrastructure to match the requirements of their existing laws. Such a step could improve environmental performance immediately, and is obviously preferable to weakening existing laws.

It must be remembered that under ISO 14001, no proof of actual compliance is actually required for an organization to obtain registration. ISO 14001 requires only evidence of working processes that are designed to maintain compliance. It is certainly a great desire and expectation that, over time, efforts to implement such processes will lead to more consistent compliance and more supportive infrastructures where they are needed.

Regulatory and Legal Implications in the United States

There is growing interest in the United States about using ISO 14001 for regulatory compliance and enforcement programs. While the U.S. Environmental Protection Agency (EPA) and the U.S. Department of Justice (DOJ) have not taken official positions on its use, there is some interest from both government bodies, and agency representatives have held preliminary discussions with leaders of the U.S. Technical Advisory Group (TAG).

Official positions from these authorities are not expected before the standards are finalized and judged to be successful. To a significant extent, that success will depend on the integrity and reliability of the third-party conformity assessment system. Government authorities will want some evidence or justification for placing their reliance on ISO 14001 registration. Such evidence must cover the accreditation and registration processes, including the rigor of third-party assessment, the independence of auditors, and the use of appropriate professional safeguards similar to those used in financial audits.

Regulators in the United States will have to consider many factors as they decide how to weave ISO 14001 into compliance programs. An organization that has been registered to ISO 14001 will have demonstrated its good-faith, voluntary efforts to better manage its environmental responsibilities and maintain compliance with applicable laws and regulations. In addition, a certified organization will have taken steps to inculcate a sense of responsibility and an environmentally conscious culture in its employees. Presumably, such an organization merits consideration from the regulators and deserves credit for its efforts. Credit could come in the form of expedited permitting, less frequent agency audits, or other means. These incentives would motivate organizations to establish an effective EMS, with the goal of continual improvement of the system, and then to become registered to ISO 14001.

Regulators are also likely to consider registration in their exercise of prosecutorial and sentencing discretion. Both the EPA and the DOJ use guidelines to weigh evidence of environmental management systems for these purposes. It is reasonable to expect that ISO 14001 may become the model used, particularly since it covers a wider number of management elements than the DOJ guidelines and, most important, encourages third-party audits for certification. It is important, however, that regulators not use the absence of ISO 14001 as a penalty against an organization. Since ISO 14001 is a voluntary standard, the only appropriate approach is to reward those who use it, not to punish those who do not. Care must also be taken not to depreciate the significance of ISO 14001 by giving it an insignificant role in voluntary or regulatory schemes. ISO 14001 transcends the limited achievements of regulatory compliance and should be justly valued and accorded the recognition it deserves.

Further, it can be expected that some courts of law will use ISO 14001 as a measure of standard commercial practice or reasonable care. Showing conformance to the elements of ISO 14001 could be very advantageous in civil and criminal liability suits. Indeed, evidence of registration to ISO 14001 is likely to have standing in a court of law, and could be used as a test to determine if an organization is practicing sound environmental management. Again, the difficulty here is to avoid punishing those who have not implemented ISO 14001. Punishment is certainly not the intent, and we should remain watchful to make sure the standard is not used in that way.

Equalizing of International Regulations

Over time, ISO 14001 will be a force for equalization of environmental regulations between countries. Although this may take many years to accomplish, the authors believe that implementation of ISO 14001 will ultimately pressure countries to harmonize their environmental laws.

As organizations around the world begin developing and implementing EMS programs that conform to ISO 14001, their abilities to undertake more sophisticated environmental protection strategies will increase. Just as the implementation of individual elements of ISO 14001 increases an organization's overall environmental awareness and, consequently, its environmental care, so it follows that as an EMS continues to improve, the protection capacity of the organization will be enhanced. As this happens with more and more organizations, government leaders may actually see less resistance to reasonable and cost-effective environmental protection measures. Thus, as ISO 14001 helps organizations become more sophisticated in environmental protection, it lays the groundwork for governments to create legislation that is more protective of the environment.

In addition, once ISO 14001 is implemented, compliance to national laws will be improved, since it is a requirement in the EMS standard for an organization to have knowledge of and to follow existing country laws. As regulators find that compliance is increasing, there will be greater impetus to continue the evolution and reformation of their country's environmental laws.

An international accreditation and registration system will also serve to spotlight the relative status of national capabilities, including legal frameworks and enforcement programs. Over time, ISO 14001 registrars will increase their expertise in comparing environmental requirements around the world. As ISO 14001 proliferates, the strengths and weaknesses of national regulatory schemes will become apparent. It is reasonable to assume that certain countries will feel compelled to bring their regulations to a higher level. In particular, countries that have the technical infrastructure for managing pollution and waste (e.g., hazardous waste management units, recycling facilities, and abatement control systems) will come under subtle pressure to upgrade their legal structures.

Certainly, no one in the regulated community wants ISO 14001 to become an engine for more regulation around the world. To the contrary, the desire is to promote voluntary management systems which have benefits far in excess of those derived from mere compliance with regulations and which, over time, can supplant the command-and-control model. That is the ultimate aspiration. In the interim, however, the management standards will coexist with country laws and regulations which, for now, are still the major incentive for many organizations.

Third-Party Auditing

Because of ISO 14001, more third-party environmental audits will occur to meet the requirements for registration. This raises a number of questions, since third-party audits have been controversial in some parts of the world for years. Questions of what can be considered privileged information, what must be disclosed, and what enforcement actions are allowed on the basis of the findings of a self-audit have been hotly debated and largely left unanswered for years.

Recently, EPA drafted a proposed regulation stating that self-audit findings would not be requested during routine facility inspections; however, the agency would request these audit findings when responding to complaints or when finding significant noncompliance issues during an inspection. The proposal went on to say that self-inspections or third-party inspections would allow for mitigation of penalties for violations disclosed to EPA or otherwise discovered by the agency.

Fortunately, the third-party audits that are required for registration to ISO 14001 are audits of the EMS and not of actual compliance.

Nonetheless, there is still concern in some countries that the ISO 14001 audit findings, once seized, could be used against an organization or facility in a court of law. There is a fear that the enforcement authority could use deficiencies found in the EMS to underscore claims of negligence or other grievances.

One safeguard that the drafters incorporated in ISO 14001 is the requirement for objectivity in auditing. Auditors are to audit to preestablished criteria, which exclude auditing of regulatory compliance. In addition, auditors are not to make a determination of the continuing suitability and effectiveness of the EMS that they are auditing. The determination of continuing suitability and effectiveness of the system is left to the organization's management to determine.

The auditor's job, then, is to assess the adequacy of the process used by management to make such a determination. These types of restrictions have been engineered into the ISO 14000 standards so that management can feel comfortable conducting EMS audits. Management is, of course, free to arrange for compliance audits as it chooses. For the time being, however, these are not required for an ISO 14001 EMS audit. As organizations gain experience and confidence with EMS audits, they will relax their guard and begin to consider the benefits of third-party compliance audits as well.

Environmental Reporting

An organization may implement ISO 14001 to show commitment to environmental protection, and to enhance its public posture on environmental issues. Implementation and registration to the standard will mean that the organization has a viable environmental management system and that the organization is serious about its environmental responsibilities. The organization's image to employees, stockholders, regulators, and the general public will be one of environmental leadership.

A likely result of this improved image will be increased environmental reporting to the public. It is true that, in many parts of the world, environmental status reports already are being published by an increasing number of organizations, particularly large organizations, such as companies on the Fortune 500 list. However, the greater majority of business organizations continue to view environmental reporting as risky and potentially unrewarding.

There are several reasons that businesses take this viewpoint. One is that many smaller businesses do not have a coherent, articulated environmental policy. In addition, many do not have a management system to guide and structure their environmental efforts. Without an environmental roadmap, these organizations have no goals, objectives, targets, benchmarks, or other underpinnings necessary to ensure an adequate level of environmental care. Thus, they lack the confidence (and the facts) to publish information about their environmental positions and progress. Further, scrutiny from the public is intimidating for any business organization, even when its environmental record is without flaw. The fear of drawing unwanted attention from employees, regulators, stockholders, and other interested parties through voluntary public reporting makes the option seem undesirable.

ISO 14001 will help organizations to better position themselves to make voluntary environmental reports. A working and effective EMS induces the discipline and focus needed to establish a suitable environmental policy, with objectives and targets, management reviews of progress, and other key elements. Once those elements are in place, it is only a question of time before the organization will have the confidence and the data to declare publicly its commitment to environmental care.

At present, environmental reporting is optional under the ISO 14001 specification, so no organization will be forced to make public reports in order to become certified. Nonetheless, as organizations begin to have positive results in environmental management, they will be proud of their successes and will want to report their progress to the public. In turn, as the popularity of these reports grows, the impetus toward greater use of environmental systems is sure to grow in parallel.

4
Conformity
Assessment

*It may well be that we'll have many member
countries and many aspiring business
enterprises that would like to start down this
road but are far from ready to face
certification, yet...are perfectly ready and
willing to try to develop their own
management system relevant to the state of
their development.*

GEORGE CONNELL
Chairman, TC 207

Conformity assessment as defined by the Office of Standards and Services of the National Institute of Standards and Technology (NIST), which is part of the U.S. Department of Commerce, means "the systematic evaluation of a product, process, or service to determine the extent to which it complies with specified requirements." It is likely that many organizations will seek certification of their conformance to ISO 14001, the environmental management system (EMS) standard. Thus, conformity assessment, as such, is fated to play a significant role in the implementation of ISO 14001.

One method of asserting conformity to a standard is through "supplier declaration of conformance"—which, simply put, means that an organization declares itself in conformance with the requirements contained in a standard. When an organization uses this method for claiming conformity, no additional verification measures are needed. In some situations, such supplier declaration of conformity may be accepted by other parties. In most cases, however, a second method of proving conformance—that of using third-party audits for conformity assessments—will be preferred. Third-party conformity assessments have played a major role in the ISO 9000 quality management system, and it is expected that similar assessments will have equal importance in the ISO 14001 environmental management system.

Elements of Conformity Assessment

Registration Using a Third-Party Audit: The ISO 9000 Experience

Conformity assessment of an ISO 9000 system typically results in registration of conformity (termed *certification* outside the United States). Registration comes after the initial assessment or periodic reevaluation audit of the supplier's quality management system by a third party, known as the quality system registrar. When a supplier's system conforms to the registrar's interpretation of ISO 9000, the registrar issues that supplier a certificate of registration. This registration process typically leads to international acceptance of the fact that the supplier's quality system conforms to the requirements of the ISO 9000 standard.

Why Use a Third Party?

In some cases, an organization may need an independent third party to verify and give credence to the fact that the organization is in conformance to a standard. Today, the public expects companies to perform financial audits using an independent third-party auditor. These third-party audits provide assurance to second parties that the financial claims of companies are accurate and a fair representation of their financial condition. Likewise, in many countries, customers expect suppliers to use third-party auditors to provide assurance of conformance to the ISO 9000 quality management standards.

There are several sets of stakeholders that have an interest in how an organization manages its environmental responsibilities. These include customers, employees, community groups, stockholders, and government agencies. To achieve maximum credibility with these parties, third-party audits which result in registration to ISO 14001 may become necessary.

Selection of a Registrar

The selection of a properly accredited registrar is very important for any organization seeking registration. The value of any registration is worth only as much as the merit ascribed to it by other parties. This is very similar to the value given to university degrees. Generally, degrees received from unaccredited colleges have virtually no standing or recognition in the marketplace. In conformity assessment, entities that set themselves up as registrars must be accepted and recognized as upstanding, competent, and credible if their registrations are to be valued in the marketplace. Obviously, no organization would seek registration from a registrar whose certificates had no standing or value with other parties. Registrars must prove that they follow internationally accepted guidelines and protocols, that they have procedures in place to carry out their responsibilities, and that they are unbiased, objective, and professional. The mechanism used to ensure that registrars merit their standing and privilege is the registrar accreditation process.

Accreditation is the process by which an authoritative accreditor establishes that a laboratory, a certification body, or a quality system registrar or any other type of registrar meets all preestablished requirements and is competent to conduct registration audits. Accreditation authorities are normally established at the national level. Their makeup and standing are established either by government fiat or, as in the United States, by private-sector action that brings recognized authorities into a body that is then accorded accreditor status by relevant stakeholders.

ISO, through its Conformity Assessment Committee, has published guides that are specific to the registrar accreditation process. The accreditation body uses these guides when it investigates the qualifications of a registrar that is seeking accreditation.

Certification of Auditors

In addition to accrediting registrars, accreditation bodies (not necessarily the same one that provides registrar accreditation) may certify auditors. Certification is defined in ISO Guide 2-1991 as "a procedure by which a third party gives written assurance that a product, process, or service conforms to specified requirements."

To become accredited, a registrar must have in its employ auditors who are competent to perform their duties, and these auditors are often evaluated and certified as meeting necessary qualifications.

Thus, most registrars will require auditors to be certified by an authoritative body before they will hire them. Certification of auditors helps the registrar obtain accreditation and ensures that only those who pass appropriate tests and/or have the proper education, training, and experience become assessors.

Accreditation of Educational Offerings

Training courses are also accredited by an accreditation body. Training providers offer courses to educate and train auditors so they can qualify as certified assessors. For example, when auditing to the ISO 9000 requirements, auditors need knowledge of how to develop and implement a quality management system, as well as of specific protocols, auditing procedures, and other information. For ISO 14000 auditing, the auditors need knowledge of how to develop and implement an EMS, as well as knowledge related to environmental issues, auditing procedures, and regulatory mandates.

Interdependence Within Conformity Assessment

While each of the above-described elements of a conformity assessment system is nominally independent of the others, the reality is that they are interdependent for the purpose of creating a system that achieves the desired outcome—registration that is recognized and valued internationally. The conformity assessment system rests on at least four legs, each of which must be as strong as the others for the system to be balanced and effective.

First, the auditors must be trained and knowledgeable if their assessments are to have credibility. Second, the courses they attend for education and training must have the proper content and appropriate educational rigor; otherwise, auditors will not acquire the requisite knowledge. Third, registrars must follow international guides and professional protocols to safeguard the integrity and standing of registrations. Fourth, accreditors must follow appropriate guides and protocols to ensure that no part of the system receives accreditation before it satisfies all qualitative and procedural requirements.

The integrity of the overall system is only as good as its weakest part. At the international level, any weakness will become known and this may well compromise a national conformity assessment system.

Establishing a Conformity Assessment System

A country that elects not to establish its own conformity assessment system can rely on existing, recognized registrars from other countries to perform registrations in that country. For instance, the United States could have British, Dutch, Canadian, or other country-accredited registrars perform assessments for U.S. organizations. For a country the size of the United States, however, this is neither practical nor desirable. In addition, it is not advisable—from either a business perspective or a national vantage point—for any country to be so dependent for its conformity assessment needs.

Establishing an Accreditation Body

Most countries will face several challenges when establishing their conformity assessment systems for use with the ISO 14000 standard. The first of these is the establishment of an accreditation body. The accreditation body is the keystone of the conformity assessment process, and as of late 1995 most countries had not yet established accreditation bodies for conformity assessment to ISO 14000.

ISO 9000 Accreditation Bodies in the United Kingdom and the United States. When the ISO 9000 conformity assessment process began, many countries endorsed either ISO 9000 or an ISO 9000 equivalent standard. As a result, these countries went about establishing government-sanctioned accreditation bodies for ISO 9000, if these bodies were not already in existence.

A widely known example of a government-sanctioned accreditation body was Great Britain's National Accreditation Council for Certifying Bodies (NACCB), sponsored by the British government's Department of Trade and Industry. NACCB accredited registrars to provide quality system registration for a particular market or range of markets. The NACCB has now been merged with the National Measurement Accreditation Service to form a new, private-sector body to be known as the United Kingdom Accreditation Service (UKAS).

In addition to the UKAS, Great Britain has a separate government-sponsored body for assessor training and accreditation, called the Registration Board for Assessors (RBA). Great Britain is the first country to set up this type of separate assessor training and accreditation board, and it is considered a prototype that some other countries plan to emulate.

Although the U.S. government has an interest in and is committed to quality systems management, it has not endorsed ISO 9000 or an equivalent standard, and it has not set up a government-sanctioned accreditation body to accredit registrars, certify assessors, or accredit training courses. As a result, any organization in the United States could potentially identify itself as a quality system registrar and issue ISO 9000 registration certificates. Lack of government sanction could be seen as a vulnerability, since the checks and balance needed in the conformity assessment process could theoretically be bypassed when firms "self-declare" their own qualification to conduct audits and issue certifications.

To counterbalance this potentially undesirable situation for conformity assessment in the United States, the American National Standards Institute (ANSI) joined forces with the American Society for Quality Control (ASQC). In a joint venture, ASQC and ANSI published the ISO 9000-9004 series verbatim as a standard, known as ASQC/ANSI Q90, and initiated plans to establish an accreditation board for ISO 9000. Ultimately, it was decided that ASQC would structure and run a subsidiary to perform that function. ASQC was chosen because of its prominence in the quality arena. ASQC accepted the task and assigned it to its new subsidiary, the Registrar Accreditation Board (RAB). RAB carries the support of both ANSI and ASQC, and since its inception has accredited some four dozen private American firms as registrars for the ISO 9000 standards. Additionally, RAB accredits courses and certifies auditors.

Because of the excellent reputations of both ANSI and ASQC, there was little or no resistance to RAB accreditation in the United States. Today, many (although not all) of the firms accredited by RAB enjoy worldwide acceptance, even though RAB is not government-sanctioned. In addition, some of the more prominent U.S. registrars have reached mutual recognition agreements with accredited registrars from other countries, providing them with added recognition and acceptability on the international scene.

ISO 14000 Accreditation in the United States.   As was the case for ISO 9000, there are no plans for a government-sanctioned accreditation body to accredit U.S. organizations as registrars for the environmental management standards. Further, there is not a single equivalent body to ASQC in the environmental arena that has similar recognition and respect to deal with environmental management assessment, as ASQC did with ISO 9000.

Although the U.S. Environmental Protection Agency (EPA) has much knowledge in environmental matters, most organizations do not think that EPA should become the accreditation body, because of its long-standing role as environmental enforcer. There is a fear that EPA would turn the accreditation program into an enforcement command-and-control program. There are also objections to giving exclusivity over the conformity assessment system to an environmental or industry group. Clearly, the U.S. system will need to accommodate the interests and views of all major stakeholders.

The outlines of a U.S. system are beginning to take shape with major roles envisioned for ANSI, RAB, the Environmental Auditing Roundtable (EAR), and perhaps the American Society of Testing and Materials (ASTM). Major stakeholders, such as EPA, state agencies, national environmental groups, and industry associations (e.g., Business Roundtable, Chemical Manufacturers' Association) may have significant roles in advisory and governing boards as well.

Registration Issues

Existing registrars that have been performing registration to the ISO 9000 quality system standards have indicated their desire to perform registration to ISO 14000. This comes as no surprise, since accredited registrars are in business for just one purpose—to perform registration services. Still, there are qualification issues that must be considered when a registrar moves from the quality arena to the environmental arena.

There are certain protocols of ISO that must be followed no matter what kind of registration is performed; thus, people who perform the ISO 9000 registrations have a good working knowledge of protocols and safeguards that must be followed in the registration process. What these quality assessors will initially lack is expertise in the environmental area. These skills will either be learned or hired to ensure that registration to ISO 14001 is performed effectively and competently. Ultimately, personnel performing registrations to the ISO 14001 standards will need an environmental background, training in ISO registration protocols, understanding of the scope and content of the environmental management standard, and certification of their competence from an accreditation body. In addition to existing registrars that wish to expand their scope to include EMS registration, there will likely be new firms seeking accreditation as registrars to perform EMS registrations exclusively.

Assessor Certification

There are already several organizations in the United States that perform environmental assessor, or auditor, certification. In the quality area, there were no preexisting certification schemes when RAB took on that task early in the process. In the environmental area, however, the situation is somewhat more complex, since these preexisting entities will want to be brought into the system. Understandably, they do not want to see their roles (and livelihood) usurped by a newly created body. Consideration is being given to the concept of having the accreditation body accredit existing organizations to perform the assessor certifications of individual auditors. The details of this approach have yet to be worked out.

Government Role in Conformity Assessment

One peculiarity of conformity assessment in the United States is the relative lack of government involvement compared with other countries. In other countries where conformity assessment is done, government involvement is a key ingredient of the system and the accreditation body is government-sanctioned. Thus, the United States is in a somewhat awkward position in terms of the international standing of its conformity assessment process. In addition, some countries have regulations that require certain registrations or certifications to be given by a government-recognized entity. Such recognition is clearly not possible in the United States, since the government is mostly not involved. This situation sometimes impinges on the acceptability of U.S. goods or services in certain markets and may require additional steps from U.S. exporters to qualify their products or services.

NIST is trying to correct the problem by putting in place a system which would fulfill the role of government recognition for U.S. accreditation bodies. Under the proposed system, if an accreditation body meets certain requirements, NIST will grant recognition to that body. Although the work is going forward, there is no specified date when NIST is expected to begin providing this type of recognition. Since ISO 14000 standards are voluntary, it is assumed that the lack of government recognition will not detract from the acceptability and credibility of the U.S. system.

Applications of Conformity Assessment Within ISO 14000

There are numerous documents within the ISO 14000 series that serve different purposes. There are documents that address the EMS, as well as subsystems and tools for environmental labeling (EL), environmental auditing (EA), environmental performance evaluation (EPE), life cycle assessment (LCA), and environmental aspects in product standards (EAPS). Of these numerous standards, only two are readily adaptable to the conformity assessment process, The first is ISO 14001, the EMS standard, and the other is ISO 14024, the EL standard for third-party (seal) programs.

Conformity Assessment As It Relates to ISO 14001

The ISO 14001 specification document requires an organization to implement an EMS that can consistently satisfy its environmental responsibilities, whether they be legal, societal, or self-imposed. It is the organization's EMS that is subject to conformity assessment.

To ensure that conformity assessment systems are comparable from one country to another, there needs to be consistency among their respective parts. Accreditors must follow the same rules and approach accreditation of registrars, assessors, and training providers with the same rigor. Registrars, in turn, must follow ISO protocols and procedures, provide the necessary safeguards, and employ competent auditors.

Further, comparability of conformity assessment is dependent on interpretations of the ISO 14001 specification document, and these interpretations must be consistent from one registrar to another and from one country to another. Areas or elements within ISO 14001 that might allow for varied interpretations or misinterpretations are presented in Fig. 4-1.

In addition to those shown in the figure, the regulatory arena is another area in which disparities in the registration process will likely occur. ISO 14001 requires an organization to have knowledge of the country laws and to establish processes to comply with those laws. It does not prescribe a benchmark or standard for what level of environmental protection is adequate. Since environmental laws vary widely from country to country, what is expected of organizations in terms of environmental protection will likewise vary widely from country to country. The ISO 14001 conformity assessment will not (and should not) address this disparity in national laws. Nonetheless, disparities exist and should be mentioned.

Along the same lines, enforcement attitudes are different from country to country, and this will again lead to inconsistencies in registrations. It is hoped that Technical Committee (TC) 207 of ISO will address these issues through the development of guidance documents that will serve as a basis for comparability in conformity assessment.

Potential Areas for Misinterpretation of ISO 14001

✓ Assessment of an organization's environmental performance is not specified and should not be assessed, but may be considered as an indicator of whether the management system is working.

✓ Interpretations of "environmental aspects" may differ from organization to organization. The listing of environmental aspects as they relate to an organization's activities, products, and services may vary greatly worldwide for organizations that are largely the same.

✓ There is a requirement for EMS audits, which are not the same as compliance audits. Compliance audits are not required in ISO 14001, but may be used to satisfy the requirement to implement processes to maintain compliance to regulations.

✓ There is a requirement for maximum auditor objectivity, based on established audit criteria, during registration assessments and periodic audits. Subjectivity is discouraged in EMS audits.

✓ The assessment should focus on process and not on output.

✓ The management review process should be assessed to determine whether the process is sufficient to allow upper management to determine the environmental management system's suitability and effectiveness. The auditor is not required to determine the suitability and effectiveness of the management system, only the efficacy of the review process to make that determination.

✓ Continual improvement of the EMS is specified in the standard and should be assessed. Continual improvement of environmental performance is not specified and is not a criterion.

✓ The concept of prevention of pollution is not well defined at an international level. The requirement that the organization consider prevention of pollution does not mean that the organization has to ensure prevention of pollution if it is uneconomical, unfeasible, impractical, or otherwise not achievable. It simply has to be considered as the first option.

Figure 4-1. Areas of ISO 14001 that might have varied interpretations.

Conformity Assessment As It Relates to ISO 14024

Although not currently completed, the intent of the ISO 14024 EL document, as described in Chap. 2, is to establish the principles and protocols that third-party labeling programs should follow when developing environmental criteria for a particular product. At present, there are approximately two dozen third-party labeling programs worldwide. A reasonable approach to accreditation of these programs is to establish a single international authoritative body to accredit all of them worldwide. This would ensure that the same accreditation criteria were applied to all labeling programs and that assessments are performed in a consistent manner. The result would promote worldwide comparability in EL criteria for products.

The Role of the Conformity Assessment Committee

The Conformity Assessment Committee (CASCO) of ISO was formed in 1985. Its predecessor, the Certification Committee (CERTICO) of ISO, was concerned primarily with the principles and practices of product certification. CASCO's mission was expanded from this and includes objectives shown in Fig. 4-2. Excluded from the mission is the task of providing guidance for the qualifying of personnel.

Over the past decade, CASCO has developed guides specific to two key areas. These are assessment and registration of quality systems

---

CASCO's Mission

✓ To study the means of assessing the conformity of products, processes, services, and quality systems to appropriate standards or technical specifications.

✓ To prepare international guides relating to the testing and certification of products, processes, and services, and to the assessment of quality systems, testing laboratories, inspection bodies, and certification bodies and their operation and acceptance.

✓ To promote mutual recognition and acceptance of national and regional conformity assessment systems, and the appropriate use of international standards for testing, inspection, certification, assessment, and related purposes.

Figure 4-2. The mission of ISO's Conformity Assessment Committee.

and requirements for assessment and accreditation of certification or registration bodies.

International Recognition Program for Quality System Registrations

CASCO was recently asked by ISO to study the feasibility of operating an international system for the recognition of quality system registrations to the ISO 9000 standards. The study was performed by an ad hoc group of ISO representatives made up of suppliers, certifiers, and accreditation bodies. The study concluded that the establishment of such a system was indeed feasible, and another group—the Quality System Assessment Recognition (QSAR) group—was formed. QSAR may eventually turn into an operating system.

ISO/CASCO has already developed two documents that are likely to be adopted by QSAR—ISO/IEC Guide 61: "General Requirements for Assessment and Accreditation of Certification/Registration Bodies," and ISO/IEC Guide 62: "General Requirements for Bodies Operating Assessment and Certification/Registration of Quality Systems." Once adopted, these should be accepted quickly by those bodies expected to participate in QSAR. In addition, CASCO is publishing a guidance document for mutual recognition agreements among accreditation bodies.

CASCO's Role in ISO 14000

It is likely that ISO/IEC Guides 61 and 62 will be the basis for ISO 14000 accreditation and registration. There is also some speculation that a QSAR-like international system for mutual recognition of conformity assessment might be beneficial in the environmental area. The International Accreditation Forum (IAF), a loose consortium of accreditation bodies from around the world, is itself becoming a factor that may play a significant role in this area.

Since CASCO does not have the mission of providing guidance for qualifying assessors, it does not provide guidance for auditor certification programs. There is considerable potential for great variability in assessor qualifications from one country to another. ISO 14012 lays out some basic requirements for assessor qualifications. However, additional work is needed in this area to ensure the basic competency of environmental assessors in all countries.

PART 2

Implementing ISO 14001

5

Environmental Policy

The ISO 14000 series presents a great way to manage the achievement of any environmental responsibility objective, including the chemical industry's Responsible Care® initiative.

JOHN MASTER
Senior Adviser, Chemical Manufacturers Association, and Chairman, U.S. Subtag on Environmental Performance Evaluation

Over the last 25 years, protection of human health and environmental responsibility have been priority concerns for industrial nations around the world. In 1972, the first United Nations (UN) Conference on the Human Environment was held in Stockholm, Sweden. That conference was the first initiative toward global environmental management, and provided guiding principles and an action plan for industrialized countries for the next two decades.

Guiding principles for environmental protection have served as a basis upon which to build an environmental policy. They provide guideposts for the actions of the organization and aspirational benchmarks in areas of importance for environmental protection. A discussion of key environmental guiding principles that have shaped U.S. and worldwide policies follows.

U.S. Guiding Principles

The United States set forth its guiding principles for environmental protection in the National Environmental Policy Act (NEPA) of 1969. Environmental principles defined in NEPA are shown in Fig. 5-1. Such principles evoke and promote a commitment to environmental protection and stewardship of the earth's natural resources.

UN Guiding Principles

The UN Conference on the Human Environment met in Stockholm on June 5–16, 1972. The ensuing report from that event set forth 26 principles "to inspire and guide the peoples of the world in the preservation and enhancement of the human environment" (United Nations, 1973). The first ten of these principles are enumerated in Appendix C. The principles are far-reaching and cover such topics as our responsibility to protect and improve the environment, resource management, economic and social development, international coordination of environmental improvement efforts, applied science and technology for environmental improvement, and education.

In June 1992—20 years later—the UN Conference on Environment and Development (UNCED) was held in Rio de Janeiro. The conference reaffirmed the work that had been done in Stockholm and found it desirable to add to the original list of principles. The resulting 27

NEPA Environmental Principles

✓ Fulfill the responsibilities of each generation as trustee of the environment for succeeding generations.

✓ Ensure for all Americans safe, healthful, productive, and aesthetically and culturally pleasing surroundings.

✓ Attain the widest range of beneficial uses of the environment without degradation, risk to health or safety, or other undesirable and unintended consequences.

✓ Enhance the quality of renewable resources and approach the maximum attainable recycling of depletable resources.

Figure 5-1. Environmental principles defined in the National Environmental Policy Act of 1969.

guiding principles—ten of which are presented in Appendix C— emphasize sustainable development and international cooperation toward common environmental goals.

Business-Initiated Guiding Principles

Principles Established by the Chemical Manufacturers Association

Various business groups, including international organizations, have also set forth broad statements of environmental principles. One prominent U.S. association focusing on environmental management issues is the Chemical Manufacturers Association (CMA). This industry group has followed the Responsible Care® initiative, which addresses six aspects of environmental management:

- Community awareness and emergency response (CAER)
- Process safety
- Pollution prevention
- Distribution
- Employee health and safety
- Product stewardship

The Responsible Care® Program is itself based on guiding principles, which are presented in Fig. 5-2. Under Responsible Care®, member companies have committed to incorporate CMA's guiding principles into their policies and activities. Further, the companies have committed to implement proactive approaches into their environmental management processes. These include the formation of public advisory panels, self-evaluations, performance measurement, management systems verification, and various other commitments.

The American Petroleum Institute has developed a program similar to Responsible Care®, called Strategies for Today's Environmental Partnership (STEP). Numerous other industrial and nonindustrial organizations have included environmental principles in their charters. One of the more prominent of these was enacted by the International Chamber of Commerce, headquartered in Paris, France.

Guiding Principles of CMA's Responsible Care® Initiative

✓ To recognize and respond to community concerns about chemicals and our operations.

✓ To develop and produce chemicals that can be manufactured, transported, used, and disposed of safely.

✓ To make health, safety, and environmental considerations a priority in our planning for all existing and new products and processes.

✓ To report promptly to officials, employees, customers, and the public, information on chemical-related health or environmental hazards and to recommend protective measures.

✓ To counsel customers on the safe use, transportation, and disposal of chemical products.

✓ To operate our plants and facilities in a manner that protects the environment and the health and safety of our employees and the public.

✓ To extend knowledge by conducting or supporting research on the health, safety, and environmental effects of our products, processes, and waste materials.

✓ To work with others to resolve problems created by past handling and disposal of hazardous substances.

✓ To participate with government and others in creating responsible laws, regulations, and standards to safeguard the community, workplace, and environment.

Figure 5-2. Guiding principles of CMA's Responsible Care® Initiative (*reprinted by permission of the Chemical Manufacturers Association*).

International Chamber of Commerce Charter for Sustainable Development

The Business Charter for Sustainable Development was prepared by the International Chamber of Commerce (ICC) in 1990 for launching at the Second World Industry Conference on Environmental Management (WICEMII) in April 1991. It provides a basic framework for environmental protection by individual corporations and business organizations throughout the world. It incorporates the best elements of various other charters and policy statements published in the late 1980s.

The ICC Charter consists of 16 principles for environmental management which can readily be adapted, in whole or in part, by organizations when structuring their environmental policies. The principles mirror most of the management elements in ISO 14001 and, as such, can be valuable when planning the implementation of an environmental management system (EMS). Principles of the ICC Charter are listed in Appendix D.

Business-Initiated Principles As They Relate to ISO 14001

It is important to note, for ISO 14001 registration purposes, that organizations which have endorsed the ICC Charter will require proof that the principles are being fulfilled through specific activities or programs. This is equally true for Responsible Care®, STEP, or any other voluntary code, charter, or program that the organization endorses or subscribes to. In effect, ISO 14001 expects organizations to take their commitments and responsibilities seriously, whether those commitments are voluntary or imposed by law.

Top Management Commitment to Environmental Policy

Definition of Environmental Policy

As specified in ISO 14001, the environmental policy is a "statement by the organization of its intentions and principles, in relation to its overall environmental performance which provides a framework for action and for the setting of its environmental objectives and targets" (Section 3.8). For purposes of this standard, an organization is any organized body or establishment—such as a business, company, government department, or nonprofit organization. Organizations with multiple operating units may opt to treat each as a separate organization under ISO 14001.

An environmental policy can be based on guiding principles, such as those discussed previously, and tailored to fit an individual organization. The policy should apply to that organization's activities, products, and services. It should reflect the organization's mission and values, and should show commitment, leadership, and direction for the organization's environmental initiatives.

Top Management Commitment

Implementing a sound environmental management system (EMS) depends on commitment from all levels of management and employees, but a commitment from top management is of utmost importance. Thus, it is not surprising that ISO 14001 calls for top management to define an organization's environmental policy.

The exact makeup of "top management" may be variously interpreted by different organizations. Some organizations may consider top management to be the chairman of the board, the president of the company, or the chief executive officer, while others may consider a senior vice president to be the top manager (e.g., the senior vice president in charge of environmental affairs for the corporation). Organizations with multiple facilities that are separately managed may consider the general manager to be the top manager of each facility. To a large extent, questions of interpretation as to "what is the right level to formulate environmental policy" and "who is top management" will be answered as implementation and conformity assessments evolve and mature. Undoubtedly, such questions will ultimately require consensus and pronouncement from Technical Committee (TC) 207.

Top management should ensure that the policy is implemented throughout the organization. The commitment from top management to sound environmental practices serves as the basis for developing and improving the EMS.

Key Elements of the Environmental Policy

Relevance

As defined in Section 4.1 of ISO 14001, an organization's environmental policy must be "relevant to the nature, scale, and environmental impacts of the organization's activities, products, and services." This wording allows an organization to customize environmental policy to fit its own needs. Such an approach is appropriate, since the specification document is to be used by all organizations around the world, whether they are in developed or developing countries, and whether they are large organizations or small organizations with activities that create significant or minor environmental impacts.

Thus, the policy can be shaped to fit the organization, and should reflect the reality of the organization's environmental situation. For instance, if an activity uses large volumes of chemicals and produces large volumes of waste, these impacts should be considered in shaping the organization's environmental policy. Likewise, if a product or service uses a significant amount of energy, then energy consumption should be an element of the policy. The policy need not be lengthy, but must clearly define an organization's environmental values and aspirations.

Commitment to Continual Improvement and to Prevention of Pollution

The policy must also include a commitment to continual improvement as well as to prevention of pollution. Continual improvement is defined in ISO 14001 as the "process of enhancing the environmental management system to achieve improvements in overall environmental performance in line with the organization's policy. *Note:* The process need not take place in all areas of activity simultaneously" (Section 3.1). The improvements are to be made to the EMS, which, once improved, may yield improvements to environmental performance. Improving the EMS allows for global, interrelated, long-term improvements. This type of approach is preferable to applying short-term performance goals, which often lead to a linear, nonintegrated solution that may only trade one problem for another. The process of continual improvement of the EMS—which itself is a continuous process—is presented in Fig. 5-3.

The principal part of continual improvement, then, is the process of enhancing the EMS. There is no obligation in ISO 14001 for an organization to continually improve its environmental performance. This distinction must be understood by the registrars in order to avoid confusion during the registration process. Examples of methods that might be used to demonstrate commitment to the prevention of pollution are shown in Fig. 5-4.

Similarly, the concept of prevention of pollution requires the organization to consider ways to prevent pollution. There is not a specific obligation to implement methods for preventing pollution if they are technically or economically impractical or otherwise not selected as viable. The requirement is for an organization to demonstrate to the registrar that methods to prevent pollution have been enumerated and evaluated before other options were considered.

Commitment to Comply with Regulations

The policy should also include a commitment to comply with relevant environmental legislation and regulations and to meet other require-

CONTINUAL IMPROVEMENT

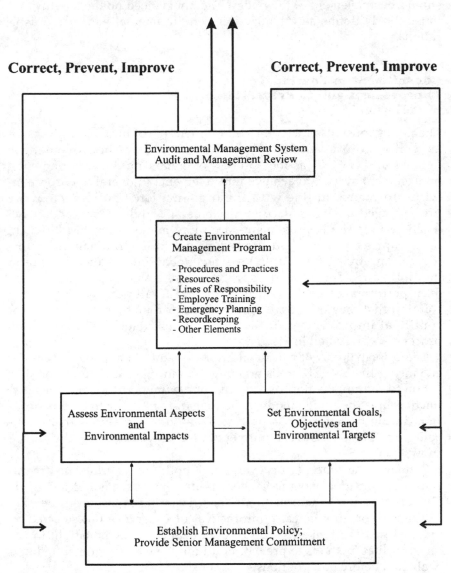

Figure 5-3. Process flow diagram for continual improvement.

Commitment to Prevention of Pollution

✓ Procedure for review of new materials to ensure that the least toxic material is selected whenever economically feasible or otherwise practicable.

✓ Procedure for identification and evaluation of best available technologies, including technical appropriateness and economic feasibility.

✓ Plans and practices for recycling of spent chemicals when practicable.

✓ Plans and practices for recycling of metals, plastics, paper, wood, and other recyclable products when practicable.

✓ Plans and practices for vehicle trip reductions through car pooling and van coordination efforts, flexible work schedules, and other feasible methods.

✓ Method for incorporating design for environment and life cycle thinking into the product development process.

✓ Method for incorporating concepts of environmentally conscious manufacturing into the manufacturing process.

Figure 5-4. Examples of methods for demonstrating commitment to prevention of pollution.

ments to which the organization subscribes. These other requirements might include commitments to voluntary programs such as the ICC Charter, Responsible Care®, and STEP. Although an organization does not have to meet all regulations and commitments to become registered to ISO 14001, the organization must have a plan in place or some other means of proving that it is working toward achieving total compliance with country laws and voluntary commitments. Examples of methods that might be used to demonstrate conformance to this requirement are presented in Fig. 5-5.

Other Elements of the Environmental Policy

Once in place, the environmental policy must provide the framework for the organization to set and review environmental objectives and targets. The setting of objectives and targets is discussed in Chap. 6.

Commitment to Compliance with Country Laws and Voluntary Commitments

✓ Procedure or matrix that specifies applicable environmental legislation and regulations.

✓ Procedure or matrix that specifies applicable voluntary commitments.

✓ Plans, with target dates, to ensure that regulations and commitments are met.

✓ Measurements of emissions or effluents that verify compliance through time.

✓ Procedure for indentifying and handling noncompliance.

✓ Corrective action plans and completion schedules for noncompliance situations.

Figure 5-5. Examples of methods for demonstrating commitment to compliance with country laws and voluntary commitments.

It is also a requirement for the environmental policy to be documented, implemented, maintained, and communicated to all employees. Documentation can take the form of a signed statement from the board of directors or the senior-level executive. Implementation can be demonstrated through the organization's instructions, objectives, targets, strategic plan, and environmental management program. Additionally, the policy must be made available to the public and interested parties. Environmental policy can be communicated at neighborhood meetings, through stockholder publications, through newspaper ads, through presentations to community groups and public officials, and by other means.

Examples of Policy Statements

Examples of elements that might be included in environmental policy statements to meet the various requirements of the ISO 14001 standards are presented in Table 5-1. This table is not all-inclusive, but can serve as a starting point for an organization's policy assessment.

Table 5-1. Examples of Elements to Include in Policy
Statements

Continual Improvement and Pollution Prevention
- Commitment to sound environmental management practices which allow
  for continued improvement within the environmental management system
- Commitment to sustainable development that protects the environment and
  has the potential to continually improve and enhance environmental performance
- Commitment to the replacement of natural resources whenever possible.
- Commitment to pollution prevention and the manufacture of products in a
  manner which reduces releases to the environment
- Commitment to life cycle thinking when developing new products and
  processes

Legislative, Regulatory, and Other Compliance
- Commitment to compliance with all environmental regulations and, to the
  extent practicable, commitment to provide environmental protection
  beyond that which is required
- Commitment to sound management of environmental aspects so as to
  reduce, to the extent possible, global environmental impacts
- Commitment to provide technological solutions that are environmentally
  sound and to provide for technical transfer of these solutions to the benefit
  of sustainable development and the environment
- Commitment to being an environmentally responsible neighbor to the
  community

Framework for Setting and Reviewing Environmental Objectives
- Commitment to development and design of products and processes in an
  environmentally conscious manner so as to reduce consumption of
  resources, including chemicals and energy
- Commitment to recycle and reuse materials to reduce waste generation
- Commitment to maintain a safe and healthful workplace for all employees

6
Planning

*Organizations should focus on improving
their environmental management systems
even before standards are published. Much of
what these standards require is simply
thorough and responsible management that is
no different from what is already practiced in
other parts of the organization.*

JOE CASCIO
Chairman, U.S. TAG

After environmental policy is set, ISO 14001 requires an organization
to develop a plan for carrying out the policy. The planning section of
the standard (Section 4.2) requires an organization to:

- Establish a procedure to identify the environmental aspects of its
  operations
- Establish a procedure to identify legal and other requirements to
  which the organization subscribes
- Establish and maintain documented environmental objectives and
  targets at each relevant function and level within the organization
- Establish and maintain an environmental program for achieving
  objectives and targets

A discussion of each of these elements is presented below.

Environmental Aspects

An important consideration when implementing ISO 14001 is the relationship among environmental aspects, environmental impacts, and

the EMS. The environment is defined in the standard as "surroundings in which an organization operates, including air, water, land, natural resources, flora, fauna, humans, and their interrelation" (Section 3.2). An added note to the definition advises that "surroundings in this context extend from within an organization to the global system." In effect, the environment is the backdrop for an organization's activities, products, and services.

As defined in ISO 14001, an environmental aspect is any "element of an organization's activities, products, and services which can interact with the environment" (Section 3.3), and a significant environmental aspect is one that "has or can have a significant environmental impact." The phraseology is open-ended, and determining what is and what is not an environmental aspect or a significant environmental aspect could be problematic for an organization. Table 6-1 provides examples of what might be considered an environmental aspect.

Finally, the standard defines environmental impact as "any change to the environment, whether adverse or beneficial, wholly or partially resulting from an organization's activities, products, or services" (Section 3.4). Again, this definition is open-ended. Examples of what might constitute an environmental impact are presented in Table 6-2.

In essence, the elements of the EMS are shaped and implemented around the organization's environmental aspects and (potential or actual) environmental impacts. The drafters of ISO 14001 intended for the EMS to ultimately address the given environmental circumstances of the organization. If an organization has no environmental aspects, there is no need to implement an EMS. If, however, an organization has

Table 6-1. Examples of
Environmental Aspects

- Waste generation
- Wastewater discharge
- Stormwater discharge
- Point source air emissions
- Fugitive air emissions
- Automobile exhaust emissions
- Chemical use operations
- Water use operations
- Energy use operations
- Use of natural resources
- Product obsolescence
- Product disposal

Table 6-2. Examples of Environmental Impacts

Impacts on Ecology
- Impacts on flora
- Impacts on fauna
- Impacts on biological diversity
- Impacts on habitat
- Impacts on landscape and natural beauty

Impacts on Natural Resources
- Impacts on agricultural land
- Impacts on forest resources
- Impacts on water supplies
- Impacts on minerals
- Impacts on marine resources
- Impacts on energy resources
- Impacts on wetlands
- Impacts on rain forests
- Impacts on wilderness

Impacts on Pollution
- Impacts on air
- Impacts on water
- Impacts on radiation levels
- Impacts on soil erosion
- Impacts on waste generation
- Impacts on contamination levels

---

numerous environmental aspects, and many of them are significant, then the organization will require an increasingly sophisticated EMS.

It will be the registrar's task to assess the process by which environmental aspects are catalogued and categorized relative to their impact. The registrar typically should not assess whether the organization has included all aspects that he or she considers significant. In certain obvious situations, however, the registrar may consider failure to include indisputable aspects in the inventory as a sign that the process for ascertaining aspects is deficient.

Legal and Other Requirements

The organization is required to identify, or catalog, legal and other requirements to which the organization subscribes that are directly applicable to its activities, products, and services. This requirement necessitates that organizations doing business in more than one country understand the environmental laws of all applicable countries.

Procedures for satisfying legal requirements must be developed and implemented, and evidence of an organization's efforts toward this end must be demonstrated during the registration process. Because of the number and complexity of regulations in existence around the world, establishing this procedure could require extensive technical knowledge, especially for multinational organizations Examples of U.S. and other country laws that might apply to a multinational organization are presented in Table 6-3.

Environmental Objectives and Targets

ISO 14001 defines an environmental objective as an "overall goal, arising from the environmental policy that an organization sets itself to achieve, and which is quantified wherever practical" (Section 3.7). Environmental targets are "detailed performance requirements, quantified wherever practicable, applicable to the organization or parts thereof, that arise from the environmental objectives and that need to be set and met in order to achieve those objectives" (Section 3.9).

The setting of objectives and targets must be consistent with the organization's environmental policy and with its commitments to prevention of pollution. This concept is somewhat ambiguous, since it might be interpreted to mean that objectives and targets must prevent pollution. In fact, the organization must prove only that it considered prevention of pollution in its activities, services, and products. The objectives and targets, in and of themselves, do not have to show evidence of pollution prevention, although in many cases they might. Examples of objectives and targets that might be set by a large manufacturing organization are presented in Table 6-4.

Environmental Management Program

An environmental management program provides a comprehensive framework for the elements necessary to achieve the company's policies, to ensure sustained conformance to environmental requirements, and to enable continual improvement. Typical elements of an environmental management program are shown in Fig. 6-1. The ISO 14001 standard is built largely around these elements.

Table 6-3. Examples of Laws That Might Apply
to a Multinational Organization

United States
- National Environmental Policy Act
- Clean Air Act
- Resource Conservation and Recovery Act
- Comprehensive Environmental Response, Compensation, and Liability Act
- Emergency Planning and Community Right-to-Know Act
- Toxic Substances Control Act
- Federal Insecticide, Fungicide, and Rodenticide Act
- Coastal Zone Management Act
- Surface Mining Control and Reclamation Act
- Federal Land Policy and Management Act
- Energy Policy and Conservation Act
- National Forest Management Act
- Wilderness Act
- Clean Water Act
- Safe Drinking Water Act
- Endangered Species Act
- Marine Mammal Protection Act
- Emergency Wetlands Resources Act
- Superfund Amendments and Reauthorization Act

Canada
- Canada Water Act and Amendments
- Fisheries Act and Amendments
- Northern Inland Waters Act and Amendments
- Clean Air Act and Amendments
- Pest Control Products Act and Amendments
- Environmental Contaminants Act and Amendments
- Ocean Dumping Control Act
- Environmental Protection Act
- Canadian Environmental Assessment Act

United Kingdom
- Batteries and Accumulators Act
- Chemicals (Hazard Information and Packaging for Supply)
- Control of Pollution Act
- Control of Substances Hazardous to Health
- Environmental Protection Act
- Planning (Hazardous Substances) Act
- Prescribed Substances Regulations

(*Continued*)

Table 6-3. Examples of Laws That Might Apply to a Multinational Organization (*Continued*)

- Public Health Act
- Water Act

France
- Pollution Control and Water Acts and Amendments
- Environmental Protection Act and Amendments
- Wastes Act
- National Conservation Act
- Marine Pollution Act
- Environmental Impact Assessment Act
- Chemical Products Act
- Environmental Protection Zone Act
- Oil Pollution Act
- Coastal Protection Act

Germany
- Disposal of Waste Act
- Traffic in DDT Act
- Federal Forest Act
- Federal Game Act
- Plant Protection Act
- Federal Nature Conservation Act and Amendments
- Federal Emissions Control Act
- Waste Avoidance and Waste Management Act
- Environmental Impact Assessment Act
- Radiological Protection Act

Sweden
- Environment Protection Act
- Marine Dumping Prohibition Act
- Products Hazardous to Health and the Environment Act
- Forest Conservation Act
- Chemical Products Act
- Building and Planning Act
- Natural Resources Act
- Environmental Damage Act

Japan
- Water Pollution Control Law and Amendments
- Soil Pollution Control Law
- Waste Disposal and Public Cleansing Law

Table 6-3. Examples of Laws That Might Apply
to a Multinational Organization (*Continued*)

- Air Pollution Control Law
- Nature Conservation Law
- Chemical Substances Control Law
- Pollution-Related Health Damage Compensation Law and Amendments
- Protection of the Ozone Layer Law
- Wildlife Preservation Law and Revisions

Table 6-4. Examples of Objectives and Targets That Might Be
Set by a Large Manufacturing Organization

Objective: Reduce Effluents and Emissions

Targets
- Evaluate and implement actions for reducing effluents and emissions
  according to the following schedule:
  33% reduction of hazardous waste by 1997
  50% reduction of hazardous waste by 2000
  33% reduction in fugitive air emissions by 1997
  50% reduction in fugitive air emission by 2000
  20% reduction in wastewater discharges by 1997
  25% reduction in chemical use by 1997
- Evaluate feasibility of on-site recycling and distillation units for solvent and
  other chemical purification and reuse by year-end 1996

**Objective: Increase Material and Product Recycling and Use of Recycled
Products**

Targets
- Evaluate recycling vendors and establish contracts per the following schedule:
  Paper, March 1996
  Metals, May 1996
  Wood pallets, July 1996
  Plastic, September 1996
- Evaluate recycled products for use in office buildings per the following
  schedule:
  Recycled paper, March 1996
  Reconditioned furniture, October 1996
  Reconditioned computer equipment, December 1996
- Evaluate other opportunities for using recycled products by year-end 1996

(Continued)

Table 6-4. Examples of Objectives and Targets That Might Be Set by a Large Manufacturing Organization (*Continued*)

Objective: Reduce Unplanned Releases

Targets
- Reduce unplanned releases by 25% by 1996
- Reduce unplanned releases by 50% by 1998
- Reduce unplanned releases by 75% by 2000

Objective: Reduce Energy Consumption

Targets
- Kick off sitewide energy savings campaign by June 1996
- Establish an energy audit program by September 1996
- Evaluate energy-saving light bulbs, timers, and other equipment by September 1996
- Formulate long-term energy-saving plan by year-end 1996
- Implement plan beginning first-quarter 1997

Objective: Reduce Water Consumption

Targets
- Evaluate all water-using processes for water conservation savings by June 1996
- Develop water conservation plan by year-end 1996
- Implement plan beginning first-quarter 1997

An Environmental Management Program

✓ Management structure, responsibilities, organization, and authority

✓ Environmental management business processes

✓ Resources (people and their skills, financial resources, tools)

✓ Process for setting objectives and targets to achieve environmental policies

✓ Operating procedures and controls

✓ Training

✓ Measurement system and auditing

✓ Management review and oversight

Figure 6-1. Typical elements found in an environmental management program.

An organization's policies, its environmental aspects, and the laws it is subject to all directly influence the structure of its environmental management program. Such a program consists of action steps, schedules, resources, and responsibilities required for the organization to achieve both its stated short-term objectives and policy conformance. Tools that might be used to implement the environmental management program include documented processes, practices, procedures, employee training and awareness, and emergency planning.

Among other things, the environmental management program must designate responsibility for achieving objectives and targets, and must set forth a time frame by which they are to be achieved. There is no mention in the specification document that registration is dependent on successfully achieving objectives and targets—only that the program for achieving objectives and targets is established and implemented. Of course, as explained previously, lack of progress may be an indicator that the system is not working.

Also, there is no requirement in the specification for communicating information about objectives and targets or relative successes and failures with respect to achieving goals and time frames. In marked contrast is the European Union's Eco-Management and Audit Scheme Regulation (EMAS), in which public disclosure of such information is the major characteristic.

7

Implementation and Operation

Clearly, EPA is moving away from the "point-in-time" inspect-and-penalize approach, to one which recognizes the existence of solid corporate environmental management systems as a strong foundation for long-term quality environmental performance.
ROBERT J. KLOEPFER
*Vice President, Haley & Aldrich
Environmental Consulting Group*

A company can have the loftiest of environmental policies and goals and the best-laid plans for environmental excellence, yet run into a major environmental problem because of inadequate implementation and operation of an environmental management system (EMS). The major catastrophes of the last decade, including large oil spills and industrial explosions, exemplify this possibility. These types of situations can occur quickly, create extremely negative public reactions, and leave lasting damage to an organization's financial position and reputation. Almost all the world's recent environmental tragedies have stemmed from breakdowns in the process management system and, most commonly, from inadequate attention to some aspect of the organization's operations.

This chapter discusses techniques for integrating the EMS into an organization's operations. While not all risks can be eliminated, implementation of an adequate EMS can assist an organization in identifying actual and potential environmental impacts and environmental risks.

117

Additionally, once impacts and risks are identified, the organization can set objectives and targets, including the development of cost-effective strategies for minimizing environmental risks for selected operations.

Usually, an organization's EMS evolves iteratively. Unfortunately, many organizations are spurred into implementing environmentally sound operations only after experiencing problems. Others are well organized with respect to environmental management, but are searching for more streamlined ways of managing their environmental affairs. Still others are interested in weaving environmental thinking throughout their organizations for strategic advantage, such as product development, operations, and distribution.

Whatever the starting point, organizations can develop and implement an EMS to identify environmental aspects and impacts, set objectives and targets, evaluate environmental performance, and make operational adjustments for continual improvement over time. Under ISO 14001, implementation and operation of an organization's EMS will be evaluated on seven elements. These elements are shown in Fig. 7-1, and are discussed in detail below.

Structure and Responsibility

Organizations can vary widely in both their structure and the attendant roles played by individuals within that structure, yet have equally effective environmental management systems. The most critical elements are the support of upper management, line management, and the organization's employees.

Implementation and Operation

✓ Structure and responsibility

✓ Training, awareness, and competence

✓ Communication

✓ EMS documentation

✓ Document control

✓ Operational control

✓ Emergency preparedness and response

Figure 7-1. Elements that pertain to operation and implementation of an environmental management system.

The underlying values set by top management in the environmental policy play a crucial role in defining the organization's EMS. The organization that commits to an effective EMS, to regulatory compliance, and to prevention of pollution is on the road to environmental progress. Conversely, an organization that makes implementation of an EMS a paperwork exercise will get none of the benefits and may produce employee cynicism and less environmental care.

There are several helpful approaches to the subject of environmental responsibility:

- Distribute environmental responsibility throughout the organization, through the management team and the employees.

- Provide regular feedback to management and employees of the organization's conformance to the EMS and to progress in achieving objectives and targets, including milestones and problem areas. This "open book" management approach will help to engage employees and encourage broad-based support for environmental initiatives.

- Consider ways to broaden traditional roles to include environmental responsibilities. One Fortune 100 company routinely includes plant managers in corporate environmental facility reviews. This accomplishes two goals: it provides the plant manager's viewpoint in the review process, and it broadens the plant manager's knowledge of environmental requirements and opportunities. ISO 14001 requires management reviews, and may be used to encourage the plant manager's participation.

Training, Awareness, and Competence

ISO 14001 specifies two types of training to be provided by the organization: training for general awareness, for all employees of an organization, and training for competence to perform a given assignment.

Training may also be needed for contractors and suppliers performing work that, because of its nature, could have environmental impacts for the organization. Examples include contractors performing work on the organization's premises and suppliers performing work on their own premises that conforms to the organization's specifications or that involves materials supplied by the organization.

Employee Training

All employees or members of an organization at all relevant levels are to receive awareness training on:

- The importance of conformance with the environmental policy and procedures and with the requirements of the EMS

- The significant environmental impacts, actual or potential, of their work activities and the environmental benefits of improved personal performance

- Their roles and responsibilities in achieving conformance with the environmental policy and procedures and with the requirements of the EMS, including emergency preparedness and response requirements

- The potential consequences of departure from specific operating procedures

Additionally, those "performing tasks which can cause significant environmental impacts shall be competent on the basis of education, appropriate training, and/or experience, as required" (Section 4.3.2).

Training ties in with total quality management (TQM) concepts, enabling continuous improvement in all system elements. Training also fits in well with so-called learning organizations. Thoughtful development of training content to simplify and to integrate it into a coherent whole reduces training time, while improving its effectiveness.

Media for training can range from standard classroom delivery to videotaped or audiotaped sessions to computer-based, even multimedia, approaches. Computer-based approaches have an advantage in that they are typically interactive, easily incorporate tests for understanding, and, once developed, are inexpensive to deliver. Examples of types of training and available or suggested training techniques are presented in Table 7-1.

As with most elements required in ISO 14001, documentation of training is key. Maintaining adequate documentation, including who was trained, the training content, and the date of training, will help ensure a smooth ISO 14001 registration.

Contractor and Supplier Training

Organizations that want to protect their employees from contractors activities on their premises and that want to ensure contractor compliance with environmental requirements must use contractor controls. Typically, these controls involve some type of training, examples of which are presented in Fig. 7-2. Similarly, an organization may want to use training to ensure that suppliers meet both their environmental and product quality requirements. As an organization implements its EMS, contractor and supplier training pertaining to the EMS itself will be

Table 7-1. Examples of Types of Training to Include in ISO 14001 Training Plan and Available or Suggested Training Media

Hazardous Waste Training

RCRA Hazardous Waste Training

- Required training—Initial training must include classroom instruction and on-the-job training that teaches employees to perform their duties in a way that ensures the facility's compliance with RCRA regulation.

- Training media—Classroom training materials might include an overview of the RCRA law using overhead projector and transparencies; a videotape of proper hazardous waste management practices; a video documentary of an uncontrolled waste site; written examples of dos and don'ts that allow for student discussion. On-the-job training might include observation of the waste-handling processes; and supervised on-the-job instruction covering, at a minimum, the following areas: manifesting, labeling, waste packaging, segregation practices, inventory and tracking procedures, waste logging and inspection procedures, and procedures for internal transport of waste.

Hazardous Waste Operations and Emergency Response (HAZWOPER) Training— General Site Workers

- Required training—Initial training must include 40 hours of instruction off site and a minimum of three days actual field experience under the direct supervision of a trained, experienced, supervisor.

- Training media—Offsite (classroom) instruction might include an overview of the HAZWOPER law using an overhead projector and transparencies; a demonstration of types of monitoring equipment available for use at the job site, with a hands-on exercise; a respirator training video, with a subsequent respirator donning and fit test; demonstration of use of other personal protective equipment such as level A, B, and C suits, bunker gear, and face protection; instruction about hazards assessment, including a review of chemical information books, Department of Transportation (DOT) emergency response books, and MSDS review—followed by a written exercise that utilizes these information sources; an interactive computer training class on confined space; tabletop exercises for setting up work zones such as remediation areas, decontamination zones, and public access zones; and review of work procedures including safety procedures, operating procedures, and emergency procedures. Field training might include hands-on demonstrations of equipment used for remediation tasks, engineering controls, safety equipment, and correct work practices; mock emergency drills which utilize monitoring equipment, personal protective equipment, field tools, and communication devices; a test including a written test with multiple-choice and fill-in-the-blank questions and a demonstration of competent use of equipment and information resources.

DOT Training

- Required training—Training requirements include general awareness training, function-specific training, safety training, and OSHA or EPA training, as applicable. If an employee changes hazardous materials job functions, that employee must be trained in the new job function within 90 days.

(Continued)

Table 7-1. Examples of Types of Training to Include in ISO 14001 Training Plan and Available or Suggested Training Media (*Continued*)

Hazardous Waste Training

- Training media—Training might include an overview of DOT, International Civil Aviation Organization, and International Air Transport Association regulations using an overhead projector and transparencies; a video of safe packaging practices and demonstrations of actual packaging techniques; workbook sessions with sample problems for proper completion of shipping papers; team competition exercises that highlight particularly difficult concepts and details; hands-on exercises for package marking and labeling; a comprehensive computer-interactive test that includes shipping paper completion, package marking, and comprehension of regulations.

Prevention, Preparedness, and Emergency Response Training

OSHA Training for Process Safety Management of Highly Hazardous Chemicals
- Required training—Employees operating a process covered under CFR 1910.119 should receive training that includes an overview of the process and applicable operating procedures; emphasis must be placed on specific safety and health hazards, emergency operations including shutdown, and safe work practices applicable to the employee's job tasks.
- Training media—Classroom training might include an overview of the OSHA law using an overhead projector and transparencies; an in-depth review of the process safety analysis using as-built field drawings; a detailed, supervised walkthrough of the process; a review of safety and emergency procedures; a computer program that provides interactive information on process safety analysis techniques and release modeling; and a review of operating procedures. Field training could include hands-on operation scenarios, mock emergency drills, and demonstrations of alarms, bypasses, overflows, switch redundancies, and other process safety features.

OSHA Emergency Response Training (HAZWOPER)—Hazardous Materials Technician
- Required training—24 hours of training to ensure that emergency responders at this level can manage the incident.
- Training media—Classroom training might include a videotape that discusses hazardous materials, including chemicals that create physical and health hazards and the risks associated with them; a demonstration of dos and don'ts with respect to safe handling practices; a tabletop discussion of risks associated with a hazardous material incident; a step-by-step review of the emergency plan and evacuation procedure; a tabletop scenario wherein the students define the emergency zones and equipment needed to mitigate the release; a review of the incident command system; and a written test to demonstrate competency. Field training might include hands-on exercises of spill control; containment and/or confinement operations within the capabilities of the personnel and personal protective equipment of the unit; demonstrations of decontamination of personnel and equipment; field demonstrations of safety equipment; mock emergency drills; and critiquing of hands-on exercises and emergency drills.

(Continued)

Table 7-1. Examples of Types of Training to Include in ISO 14001 Training Plan and Available or Suggested Training Media (*Continued*)

Other Employee Training

Environmental System Operator's Training

- Required training—Training should ensure that operators can perform their duties with minimum impact to the environment. This training might apply to operators of the wastewater treatment plant, air abatement systems, and chemical distribution systems.

- Training media—Classroom training might include a video about environmental problems caused by operational errors; a review of operating procedures; and a review of troubleshooting techniques and corrective actions for nonconformance situations. Field instruction might include hands-on operations demonstrations; field review of alarms, monitoring equipment calibration procedures, bypasses, overflows, and redundant equipment; and mock drills of equipment malfunction.

Employee Environmental Awareness Training

- Required training—Training must make employees at all relevant levels aware of: the importance of conformance with the environmental policy and procedures and with the requirements of the EMS; the significant environmental impacts, actual or potential, of their work activities and the environmental benefits of improved personal performance; their roles and responsibilities in achieving conformance with the environmental policy and procedures and with the requirements of the EMS; and the potential consequences of departure from specific operating procedures.

- Training media—Classroom training might include an overview of the environmental policy, environmental program, and the ISO 14001 standard using overhead projector and transparencies; a review of the organization's environmental policy; a step-by-step review of operational procedures that apply to environmental protection; a video showing sound environmental management practices such as pollution prevention, source reduction, and recycling; a panel discussion of managers describing environmental objectives and targets for their areas; a review of the emergency plan and safety procedures; and written exercises and discussions about the individual employee's role in environmental management. Field training might include emergency drills; hands-on demonstration of alarms, shutdown devices, bypasses, and other emergency equipment; and demonstration of dos and don'ts of running the operation.

---

beneficial. A solid training program for contractors and suppliers can serve two purposes. First, it can make contractors or suppliers aware of the environmental aspects and impacts of their activities, which should lead to better environmental protection, prevention of environmental harm, and, ultimately, reduction in liability. Second, it allows contractors or suppliers to see the value of an integrated EMS.

Control of Contractor Training

✓ Training is given when security badges are issued.

✓ Handbooks with contractor procedures are sent to the contractor before the job begins.

✓ Periodic mandatory contractor training seminars are held.

✓ Training videos are supplied to contractor or are available for use on site.

✓ Unannounced inspections of job sites are performed to ensure compliance to site procedures; at a minimum, additional training is mandatory if procedures are violated.

Figure 7-2. Suggested contractor training methods.

Communication

Another key aspect of sound environmental management is communication with employees, with neighbors and other interested members of the public, and with customers. ISO 14001 specifies that procedures be in place for

- Maintaining internal communication between various functions and levels of the organization
- Receiving, documenting, and responding to relevant communication from external interested parties regarding environmental aspects and the EMS

Additionally, the organization shall consider processes for external communication on its significant environmental aspects and record its decision about whether to implement a communication process.

Various means of communication may prove useful, including:

- Regular reviews by management of the organization's environmental status
- Periodic presentations from the management team and employees on specific environmental challenges
- Open houses for employees' families, the surrounding community, and public officials
- Awareness training for employees, contractors, and suppliers
- Written communication, such as a periodic newsletter or an annual report, to one or more of the organization's stakeholders
- Use of an 800 number for obtaining feedback from consumers

This list is certainly not exhaustive, but it gives ideas for communication methods. An organization will want to tailor communications to fit its specific needs and goals.

EMS Documentation and Document Control

Say what you do and do what you say.

EMS processes and procedures need to be documented and kept current. When registering to ISO 14001, an organization will want to ensure that the registrar finds a one-to-one match between the documented process and what is actually practiced by the organization.

A major challenge is creating an effective means for deploying the latest documentation. Documentation can be either paper or electronic. Organizations that are small enough can put all relevant information into a binder and keep it in a central location. Larger organizations that are fortunate enough to have a computer network can organize and distribute up-to-date information via that means. Paper copy distribution can also work, as long as obsolete documents are removed from service. In all cases, the documentation should be dated (with the dates of revision), clearly identified and organized, and reviewed and updated on a fixed schedule.

Operational Control

General

Section 4.3.6 of ISO 14001 contains important requirements about operational control. In simple-sounding language, it mandates several far-reaching actions necessary for an organization to show conformance. The standard specifies the following:

> The organization shall identify those operations and activities that are associated with the identified significant environmental aspects in line with its policy, objectives, and targets. The organization shall plan these activities, including maintenance, in order to ensure that they are carried out under specified conditions by:
>
> (a) establishing and maintaining documented procedures to cover situations where their absence could lead to deviations from the environmental policy, and the objectives and targets
>
> (b) stipulating operation criteria in the procedures

(c) establishing and maintaining procedures related to the significant environmental aspects of goods and services used by the organization and communicating on relevant procedures and requirements to suppliers and contractors

In conjunction with identifying its significant environmental aspects, an organization must identify its operations and activities associated with those aspects. An example is air emissions (say nitrous oxides and sulfur oxides) from a power-generating facility. Another example is plating rinse waters discharged from printed circuit board manufacturing. Understanding equipment and process parameters is critical to identifying—and ultimately to minimizing—associated environmental impacts.

Once such operations and activities are identified, procedures need to be developed to cover the requirements noted in (a) through (c). Importantly, these include the significant aspects of an organization's operations, as well as communication of relevant procedures and requirements to suppliers and contractors.

The following suggestions may prove useful in developing these procedures:

- Keep the procedures simple to understand and to use.
- Decide beforehand how they will be distributed for use.
- Provide training and motivation to those responsible for carrying out the procedures.
- Create a systematic way for reviewing the procedures to keep them current and relevant to the users.

Supplier Management

An organization must ensure that its suppliers and contractors understand its requirements so that they do not unwittingly cause the organization to compromise its own EMS. This is a rather delicate section of ISO 14001 as there is no intent in the standard to impose the organization's EMS on suppliers.

ISO 14001 requires an organization to review "the environmental aspects of its activities, products, and services over which it has control or can be expected to have an influence." This does not give it license to include environmental aspects of its suppliers. The target of the environmental aspect review is the activities of the organization, not the activities of its suppliers. It would not be necessary, for purposes of ISO 14001, for an organization to

- Request information from the supplier or contractor about activities performed, chemicals used, waste generated, potential releases to the atmosphere, and other actual or potential environmental aspects and impacts experienced by them.

- Impose its own EMS on suppliers and contractors.
- Visit a supplier site to ensure that applicable legal requirements are being met.
- Demand that suppliers and contractors have a registered EMS.

Although none of these activities are required under ISO 14001, the standard does not prohibit an organization from going beyond what is required in terms of extending its realm of influence for environmental care beyond its regular activities, products, and services. These activities may be required in some country regulations or they may be incorporated into an organization's voluntary commitments.

Emergency Preparedness and Response

The procedures for operational control are the front line of defense to head off any need for emergency response. Nonetheless, preparing for an emergency is a critical part of any EMS. If an emergency occurs, an organized, competent response will help minimize any damage to human health or the environment. Emergency plans should include items presented in Table 7-2.

In addition to planning for emergency response, the organization might want to model potential releases from the plant site. For this purpose, source emission models might be used to evaluate the effects of gas jet releases, liquid jet releases, two-phase jet releases, and single- and multicomponent liquid pool evaporation. In addition, transport and dispersion models might be used to evaluate the short-term and long-term effects of a chemical release, including models that predict transport and dispersion of dense gas releases at grade, as well as releases in nonhomogeneous terrain and wind fields, and models that predict transport and dispersion of hazardous gases.

Finally, the organization will want to ensure that it documents the most effective release mitigation techniques for its operations. Examples of these techniques include prerelease controls and protection equipment, safety systems and procedures, and management activities. Techniques can include employee training, certification of operators on equipment and systems, membership in community emergency planning groups, development of an accident/incident investigation program, participation in research or conferences, development of a safety loss prevention program, and initiation of a program to improve system designs.

Table 7-2. Examples of Elements to Include in an Emergency Plan

Planning Elements
- Identification and description of areas on site that store, use, or otherwise manage hazardous substances
- Identification of neighborhoods, schools, hospitals, parks, wildlife habitats, and other sensitive areas around the facility that might be impacted by a release of hazardous substances
- Documentation of methods used on site for determining that a release of hazardous substances has occurred
- Description of methods to assess areas likely to be affected by an ongoing release
- Instructions for plan use and record of amendments, including listings of organizations and individuals receiving the plan or plan amendments, and other data about the dissemination of the plan

Concept of Operations, Direction, and Control
- Designation of a site emergency coordinator who will determine when to implement the site emergency plan
- Designation of other key individuals such as trained incident commander(s), trained emergency response personnel, hazardous materials specialists, medical personnel, security personnel, and communication liaison
- Description of communication methods to be used among responders
- Description of procedures for responders to enter and leave the incident area, including safety precautions, medical monitoring, sampling procedures, and designation of personal protective equipment
- Description of procedures to be followed by operations personnel in the event of release of a hazardous substance
- Descriptions of major methods for cleanup
- Information identifying outside assistance, such as local hazardous materials emergency response team, fire department, police, and medical assistance
- Emergency hot-line numbers and lists of names and numbers of organizations and agencies that are to be notified in the event of a release

Resource Management
- Description of emergency equipment on site and auxiliary equipment in the community
- List of personnel resources available for emergency response
- Description of the training program for site personnel

Personal Protective Measures/Evacuation Procedures
- Description of evacuation plans from buildings and from the site
- Information on precautionary evacuations of special populations and information on facilities that provide food, shelter, and medical care to relocated populations

SOURCE: NRT (1988).

8

Checking and Corrective Action

The most important concept underlying the ISO Environmental Auditing Guidelines is that an environmental audit is a verification process, whereby an audit team makes its finding based on a comparison of the audit evidence against the audit criteria. Thus, the primary function of the environmental auditor is to determine an auditee's conformance, not its environmental performance.

CORNELIUS "BUD" SMITH, JR.
*Director of Environmental Management
Services, ML Strategies, and Chairman,
Subtag on Environmental Auditing*

Subsection 4.4 of ISO 14001 addresses checking or monitoring activities related to the environmental management system (EMS) as well as means and methods for taking corrective action if deficiencies are found. Included in the section are:

- Monitoring and measuring the EMS
- Handling and investigating nonconformances
- Implementing corrective action and preventive action
- Maintaining environmental records
- Establishing and maintaining an EMS audit program

Tools available to organizations implementing this section of the ISO standards include:

- ISO 14004, "Environmental Management Systems—General Guidelines on Principles, Systems, and Supporting Techniques"
- ISO 14010–14012, "Environmental Auditing Guidelines"
- ISO 14031, "Guidelines on Environmental Performance Evaluation" (currently under development)

Monitoring and Measurement

Monitoring and measurement are required elements in an EMS. It is through monitoring and measurement that an organization can assess its progress in meeting stated environmental objectives and targets. A monitoring and measurement program is a continuous process that includes ongoing data collection and continual tracking of specified parameters. Examples of elements that could be part of a monitoring and measurement program are presented in Fig. 8-1.

A functional monitoring and measurement system should include:

- Procedures to monitor and measure on a regular basis key characteristics of operations and activities that can have a significant impact on the environment

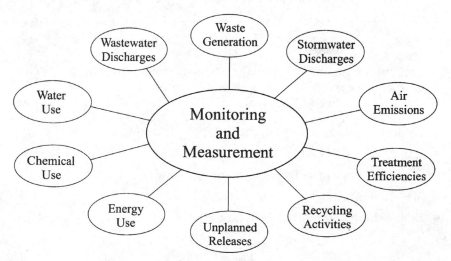

Figure 8-1. Examples of elements of a monitoring and measurement program.

- A mechanism to record information that tracks performance, relevant operational controls, and conformance with environmental objectives and targets

- A procedure to calibrate monitoring equipment, and a method for ensuring that calibration records are kept for the period of time prescribed

- A procedure for periodically evaluating compliance with environmental legislation and regulations

The following paragraphs provide insight about implementing these four elements of a monitoring and measuring program.

Procedures to Monitor and Measure Key Characteristics

The most important and challenging elements in monitoring and measurement are, first, selecting the key characteristics and, second, defining the methods of measurement. In short, the organization must determine what to monitor and how. Country laws, voluntary commitments, existing permit requirements, impacts to the environment, available monitoring equipment, contract laboratory costs, and other factors must be assessed when making decisions about characteristics to include and measurement techniques to employ. After selecting key characteristics and measuring methods, the organization can define the subsystem that provides the desired information. Examples of key characteristics and measurement methods are presented in Table 8-1.

Tracking Performance

ISO 14031 has the potential to benefit those organizations which do not already have a mature monitoring and measurement program. It will provide sample indicators and measuring methods that an organization can choose from or model for its own evaluation subsystem.

As discussed in Chap. 2, improvement of environmental performance is not a requirement to become registered to ISO 14001. Nonetheless, tracking environmental performance will allow the organization to improve its EMS, and EMS improvements will likely result in improvements in environmental performance.

Calibration of Equipment

Calibration of equipment, too, must be regularized through an established and implemented procedure. A sample equipment calibration matrix is shown in Table 8-2.

Table 8-1. Examples of Key Characteristics and Measurement
Methods

Waste Generation

Key Characteristics
- Hazardous waste, including ignitable, corrosive, reactive, toxic, or listed by a regulatory agency as hazardous
- Otherwise regulated waste, such as DOT- or state-regulated waste

Measurement Methods
- Amount generated, as related to activity index, per time period (i.e., monthly, quarterly, semiannually, or annually)
- Disposal or treatment costs, as related to activity index, per time period
- Number of waste shipments per time period
- Storage timeframes (e.g., less than 30 days, less than 60 days, or less than 90 days)

Stormwater Discharges

Key Characteristics
- Toxic Release Inventory (TRI) chemicals that might be found in stormwater, as a result of the organization's activities
- Parameters that are listed on the organization's National Pollutant Discharge Elimination System (NPDES) permit
- Oil and grease from parking lot runoff
- Other parameters that might be found in stormwater as a result of the organization's activities
- Rate of runoff from site

Measurement Methods
- Grab sampling on a periodic basis (i.e., monthly or whenever it rains) at selected outfalls; analysis of specified "indicator" parameters such as pH, chemical oxygen demand (COD), and oil and grease
- Composite sampling during first hour of rainfall event on a semiannual basis, if required under stormwater permit; analysis of TRI chemicals and other specified parameters using EPA or other acceptable method
- Flow rate sampling during all or specified sampling events

Air Emissions

Key Characteristics
- Parameters listed in the organization's Clean Air Act (CAA) permit
- Other parameters that the organization discharges into the air through abated and unabated stacks that might cause environmental impact, including ozone precursors, greenhouse gases, and toxics
- Flammable gases

Table 8-1. Examples of Key Characteristics and Measurement Methods (*Continued*)

Measurement Methods
- On-line measurement systems such as flame ionization detector (FID), photo-ionization detector (PID), combustible gas indicator (CGI), on-line mass spectrometer (MS), or other appropriate method
- Periodic stack sampling using a sampling train and an EPA or other acceptable collection and analytical method

Treatment Efficiencies

Key Characteristics
- Treatment efficiencies specified in organization's permits such as efficiencies in wastewater treatment, air abatement, incineration, and other prescribed technology efficiencies
- Treatment efficiencies required for reporting such as on the TRI report, state reports, or other reports
- Other treatment efficiencies considered important by the organization such as water purification efficiencies, chemical reprocessing efficiencies, and other efficiencies

Measurement Methods
- Mass balance
- Analytical comparisons of influents and effluents
- Use of EPA or other efficiency factors
- Manufacturer's literature estimates
- Other data

Recycling Activities

Key Characteristics
- Materials or wastes that have recycling potential

Measurement Methods
- Amount of material or waste recycled, as related to activity index, per time period
- Revenue generated from recycling activities, as related to activity index, per time period
- Disposal costs avoidance

Unplanned Releases

Key Characteristics
- Unplanned emissions or effluents exceeding permit values
- Chemical spills that breach secondary containment

(Continued)

Table 8-1. Examples of Key Characteristics and Measurement Methods (*Continued*)

- Other chemical spills
- Large-volume water spills, such as process water spills, chilled water spills, and treated wastewater spills

Measurement Methods
- Mechanisms for spill reporting and tracking
- Trends of unplanned releases per time period
- Incident response time

Energy Use

Key Characteristics
- Energy consumed

Measurement Methods
- Electric bills, based on activity index, per time period
- Periodic audits of manufacturing areas, on weekends and other times of nonproduction, to ensure equipment is turned off when not in use
- Periodic audits of office areas, on weekends and evening hours, to ensure lights and computers are turned off when not in use

Chemical Use

Key Characteristics
- Amount of high-use chemicals consumed

Measurement Methods
- Trends of use, as related to activity index, based on purchasing records per time period
- Trends of use, as related to activity index, from on-line bulk chemical monitoring system per time period
- Trends of use from other usage monitoring, based on activity index, per time period

Water Use

Key Characteristics
- Amount of city water consumed
- Amount of purified water consumed, such as water treated by filtration, reverse osmosis, or deionization
- Amount of irrigation water consumed

Table 8-1. Examples of Key Characteristics and Measurement Methods (*Continued*)

Measurement Methods
- Trends of use, as related to activity index, based on water meters per time period
- Trends of use, as related to activity index, based on water bills per time period

Wastewater Discharges

Key Characteristics
- Parameters in NPDES permit
- Parameters in state or city regulations
- Other parameters that could affect the environment, as applicable to area, such as nutrients, metals, toxics, and other parameters

Measurement Methods
- Periodic grab or composite sampling of outfalls for specified parameters, per permit requirements
- On-line monitoring of certain parameters such as pH
- Additional daily, weekly, or monthly grab or composite sampling, using EPA or other approved sampling method
- Screening sampling for batch treatment or process observation

---

Table 8-2. Sample Equipment Calibration Matrix

Wastewater Monitoring Equipment

Equipment	*Frequency*
Gas chromatograph	Daily
Mass spectrometer	Daily
pH sensor	Daily
Colorimetric spectrophotometer	Daily
Flow meter	Monthly
High-level alarm	Quarterly

Air Monitoring Equipment

Equipment	*Frequency*
On-line flame ionization detector	Weekly (self-calibration)
On-line mass spectrometer	Weekly (self-calibration)
On-line gas detector	Daily (self-calibration)
On-line carbon monoxide detector	Daily (self-calibration)
Flow meter	Monthly
High-concentration alarm	Monthly

(*Continued*)

Table 8-2. Sample Equipment Calibration Matrix (*Continued*)

Process Monitoring Equipment

Equipment	*Frequency*
Chemical add sensor	Weekly
Water meter	Weekly
Part counter	Monthly
Temperature sensor	Quarterly
Pressure sensor	Quarterly

Bulk Chemical Monitoring Equipment

Equipment	*Frequency*
Tank level	Monthly
Flow meter	Monthly
Tank overflow alarm	Quarterly
Leak detection system	Quarterly
Temperature sensor	Quarterly
Pressure sensor	Quarterly
Scales	Annually

Incident Management Equipment

Equipment	*Frequency*
Oxygen/lower explosive limit meter	Before each use
pH sensor	Before each use
Leak detection sensor	Quarterly

Water Usage Monitoring Equipment

Equipment	*Frequency*
Process flow meter	Quarterly
Cooling tower make-up flow meter	Quarterly
Cooling tower blow-down flow meter	Quarterly
Irrigation system meter	Annually
City inlet flow meter	Annually

Stormwater Monitoring Equipment

Equipment	*Frequency*
pH sensor	Weekly
Flow recorder	Quarterly

Energy Monitoring Equipment

Equipment	*Frequency*
Power use meter	Annually
Run time meter	Annually

Waste/Recyclable Materials Management Equipment

Equipment	*Frequency*
Building leak detection sensor	Quarterly
Building lower explosive limit meter	Quarterly
Scales	Annually

Periodically Evaluating Compliance with Environmental Legislation and Regulations

ISO 14001 requires an organization to establish and maintain a documented procedure for periodically evaluating compliance with relevant environmental legislation and regulations. Elements contained in such a procedure might include a list or matrix of country laws and regulations, a list or matrix of permits, and methods for evaluating compliance. Examples of methods that might be used to evaluate compliance are presented in Fig. 8-2.

The requirement does not demand that an organization be in full compliance will all laws, and the specification does not prohibit an organization from becoming registered to the ISO 14001 standard, even if it does not meet some applicable laws. Notwithstanding, the organization's environmental policy must include a commitment to comply with relevant environmental legislation and regulations, and from this requirement, the registrar can assess evidence of good-faith efforts to meet the requirement. Further, the registrar must assess if the EMS is adequate to achieve improvements needed to meet the commitment to compliance. Assessing adequacy and commitment is very different from assessing compliance itself, but it is an important concept that should be understood by organizations and registrars alike. Compliance data can be viewed, but only as an indication of how well the EMS is working.

In addition, it is important to note that the requirements do not specify that periodic evaluating of compliance with relevant laws and regulations must be made through a compliance audit. Certainly, compliance audits are a method that can be selected by the organization to evaluate compliance with laws and regulations, but there can be other acceptable methods for meeting this requirement. As long as a good-faith evaluation procedure is documented and followed, it will be adequate evidence to the registrars that this section of the standard is being met.

Nonconformance and Corrective and Preventive Action

ISO 14001 requires an organization to establish and maintain procedures for handling, investigating, and initiating corrective and preventive action for nonconformance. Additionally, responsibility and authority for all activities related to nonconformance must be defined.

Compliance with Country Laws

✓ Review of hazardous waste documentation, including permits and registrations, required agency reports, required plans, manifests and land disposal restriction notifications, training records, container and tank inspection logs, records of noncompliance release reporting, waste minimization records, documentation of incidents where the contingency plan was invoked, ground water monitoring data and reports, and other reports.

✓ Review of air emissions documentation, including construction and operating permits, emissions monitoring data and reports, new source performance test results, hours of operation, records of control system efficiency, chemical consumption tracking logs, production tracking logs, fuel oil or natural gas consumption tracking logs, fugitives monitoring records, emissions inventory, and other documents, as required.

✓ Review of wastewater/stormwater discharge documentation including wastewater discharge monitoring data and reports, spill prevention control and countermeasure plan, pretreatment compliance reports, stormwater monitoring data and reports, stormwater pollution prevention plan, whole effluent toxicity text data and reports, records on noncompliance reporting, and other documents, as required.

✓ Field inspections to ensure proper handling, labeling, and storage of waste containers; proper posting of PCB, asbestos, and hazardous waste warning signs; proper posting of other warning signs; existence of necessary procedures, inspection logs, and other documentation; proper maintenance of pollution control equipment; proper calibration of monitoring equipment; and other field items, as selected.

Figure 8-2. Examples of methods for evaluating compliance with country laws.

Nonconformance refers to deviations from the EMS and from requirements of ISO 14001 and should not be confused with noncompliance. The term *noncompliance* is used for deviations from country law and regulations. In effect, one of the requirements of the EMS is management vigilance over the continuous operation of the EMS elements. This requirement provides the feedback mechanism to correct the EMS if any part falters and needs correction. The standard further requires that steps be taken to prevent recurrence of the nonconformance.

Nonconformance includes anything that does not meet requirements, as defined by the EMS. It can include, but is not limited to, nonconformance with respect to policy, objectives, and targets; structure and responsibility; training plans; operational requirements; equipment calibration schedule; recordkeeping; document control; emergency preparedness and response procedures and scheduled drills; monitoring and measurement plans; EMS audits; and management review documentation and EMS improvements implementation.

The management system for handling nonconformance typically will include:

- Identification of the cause of the nonconformance, through root cause analysis or other methods
- Identification of options for corrective and preventive action, including the addition or modification of procedures or other controls
- Personnel training
- Implementation of a plan for selected corrective action

All corrective and preventive actions should be appropriate to the magnitude of the nonconformance and to the actual or potential environmental impact.

Records

Maintaining environmental records is a key part of an environmental management system. These records will allow an organization to demonstrate conformance to the ISO 14001 standard, as well as to track progress toward meeting objectives and targets. Examples of records that might be maintained by an organization are presented in Table 8-3. Records must be maintained so that they are readily retrievable and protected against damage, deterioration, or loss.

EMS Audit

ISO 14001 defines an environmental management system audit as "a systematic and documented verification process to objectively obtain and evaluate evidence to determine whether an organization's environmental management system conforms to the EMS audit criteria set by the organization, and to communicate the results of this process to management" (Section 3.6). The audit is required on a periodic basis, depending on the environmental importance of an organization's

Table 8-3. Examples of Environmental Records Maintained by an Organization

- Incident reports
- Complaint reports
- Contractor and supplier information
- Product information
- Process information
- Records of nonconformance and corrective and preventive action
- Procedures for emergency preparedness and response
- Management review records
- Audit records
- Agency inspection records
- Equipment inspection and calibration records
- Training records
- Monitoring records
- Records specific to environmental impacts such as waste generation and chemical usage
- Information pertaining to environmental laws

activities and the results of previous audits. The purpose of the audit is to determine whether the EMS conforms to planned arrangements for environmental management and if the EMS is properly implemented and maintained.

It is important to reemphasize that the audit is an audit of the EMS, not of environmental performance. The audit criteria pertain to the EMS and are set by the organization. The EMS audit is expected to be an objective evaluation against that preestablished criteria. Basically, the auditor determines if the EMS is in place as specified in ISO 14001 and whether it is working and attended to. An organization needs to have the audit system in place, documented, implemented, and maintained, with results reported to management, in order to achieve conformance.

It is also useful to distinguish between the EMS audit required by ISO 14001 and the registration audit conducted by registrars for certification purposes. Both can be third-party audits, but the EMS audit is conducted against preestablished criteria that the auditee has a role in developing, while the registration audit follows uniform criteria established by the conformity assessment system in a given country. Neither the EMS nor the registration audit is a compliance or performance audit. In both cases, compliance and performance data may be viewed only as indicators of whether the management system itself is working. Even here, however, the auditor needs to be cautious, since there may be other reasons for a temporary degradation of performance or higher incidence of noncompliance (e.g., when the plant receives a new mission and systems are in transition).

An effective EMS audit program should allow the organization to determine whether the EMS:

- Conforms to a planned arrangement for environmental management, including the requirements of ISO 14001
- Has been properly implemented and maintained
- Provides information on the results of the EMS audit to management for its review

The audit program procedures should specify the frequency of the audits, the audit scope, audit methodologies, and responsibilities and requirements for conducting the audits and for reporting the results. The audit scope should be limited to requirements defined by the EMS, and should not include environmental performance per se.

The audit can be conducted by personnel within the organization or by a third-party audit team. If internal personnel conduct the audit, there should be some mechanism in place to ensure objectivity.

The environmental system audit provides a snapshot in time of the effectiveness of an organization's EMS. The process is designed so that evidence, which may be qualitative or quantitative, is gathered and used to verify that audit criteria are met. Examples of methods for collecting evidence are presented in Fig. 8-3.

Methods for Collecting Evidence

✓ Interviews with personnel

✓ Examination of documents

✓ Observations of activities

✓ Observation of conditions

✓ Test data

✓ Monitoring data

✓ Other records

Figure 8-3. Examples of methods for collecting evidence in ISO 14001 registration.

The audit process compares an organization's EMS implementation with its statements of intention. Once the audit is completed, the results of the EMS audit are reported to management. Guidelines for auditing an EMS are detailed in ISO 14011. This document contains information for internal audits of the EMS system, but the information is useful for registration to the standard as well. Information that might be required during the registration audit is presented in Chap. 10.

There may be a formal written report of the EMS audit, although this is not required under the specification. Suggestions of items to include in an audit report are detailed in ISO 14011 and summarized in Table 8-4. If a written report is requested, the distribution list should be determined in advance. The audit report and related audit findings typically are managed as "confidential."

Table 8-4. Examples of Items to Include in an EMS Audit Report

Organization and Personnel
- Organization name (auditee)
- Organization structure
- Names of personnel and managers participating in the audit as auditees
- Organization name of third-party auditor (if applicable)
- Names of audit team members

Audit Protocol
- Scope, objectives, and plan of audit
- Agreed criteria of audit (include a list of reference documents against which the audit is to be conducted)
- Audit period
- Distribution list for the audit report

Audit Findings
- Identification of the confidentialities associated with the audit contents
- Summary of audit process
- Audit findings and conclusions as to EMS conformance to the EMS audit criteria
- Audit findings and conclusions as to whether the system is properly implemented and maintained
- Audit findings and conclusions as to whether the internal review process is able to ensure the continuing suitability and effectiveness of the EMS

9
Management
Review

What isn't measured quickly deteriorates.
DWIGHT D. EISENHOWER

Despite being the shortest section in ISO 14001, management review is by no means the least important. To the contrary, review by management is absolutely vital to the success of an environmental management system (EMS). Management review provides the nexus for an organization's environmental policy, long-term goals, environmental results, and continual improvement.

Management has unique and exclusive responsibilities within ISO 14001. These responsibilities are presented in Fig. 9-1. As specified in the standard, management is the leader of the organization, setting the organization's course, assessing its results, and adjusting the EMS elements to achieve short- and long-term environmental goals.

Management Review Elements

According to the ISO 14001 specification, Section 4.5, a management review is to be performed and documented at intervals determined by management to ensure that the EMS is:

- Suitable
- Adequate
- Effective

EMS Management Responsibilities

✓ Environmental policy and strategy

✓ Judgment and action on environmental management system review and audit results

✓ Judgment and action on environmental performance results

✓ Continual improvement actions, consistent with environmental policy

✓ Staffing, organizational structure, and culture

✓ Financial and technological resources

Figure 9-1. Management responsibilities in the EMS process.

The EMS must ensure that needed information is compiled to allow for a proper management review. The information includes:

- Previous management review and audit results
- Environmental objectives and targets versus performance results
- Changes in business environment that may influence policy, objectives, and targets
- New or changed legislation
- New or changed stakeholder or interested-party expectations
- Changes in applicable technology, including work processes
- Organization's financial and competitive position
- Business areas and activities
- Market preferences
- Environmental incidents, nonconformances, and corrective action

Senior management involvement is key. Since senior managers are responsible for setting environmental policy, their participation is essential to ensure sufficient feedback for continual improvement. Management involvement will also be important for obtaining registration, since it shows commitment to the environmental policy and its successful application. Management reviews can be simple or involved, informal or formal, reflecting the organization's culture.

Role of Environmental Management Staff

Environmental management staff plays a role to ensure a productive and effective management review. Several tasks can best be performed by those with environmental management responsibilities. These are delineated in Fig. 9-2. In addition, the environmental management staff provides the technical assistance to line management to improve the EMS on the basis of management assessments and directions.

Management Review Approach

As stated previously, management reviews should reflect the organization's culture and style, as well as the preferences of the individuals involved. There are many approaches that management can use to structure its reviews. In general, these will involve a combination of formal and informal methods. Formal methods include:

- Regular update and review of a given set of program and process measurements (also referred to as metrics or indicators)
- In-depth review of program and process elements, such as requirements, ownership, process flows, procedures, interfunctional dependencies, measurements, control points, and auditable elements
- Review of nonconformances, which may be included in scheduled periodic reviews, reviewed real time, or both

---

Role of Environmental Management Staff

✓ Highlight current and emerging issues

✓ Coordinate EMS audits

✓ Oversee the EMS, including underlying processes and performance indicators being used

✓ Provide guidance on environmental performance measurements to line and supporting organizations

✓ Collect, analyze, and review with management environmental performance measurements of the entire organization

---

Figure 9-2. Role of environmental staff in the EMS process.

- Framing and review of environmental policy, EMS, and strategy for continual improvement

Examples of measurements that might be shown during management reviews are presented in Figs. 9-3 to 9-5.

Executives also use informal methods to stay in touch with how things are working, or not working, at a plant site. "Management by

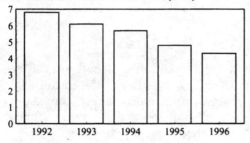

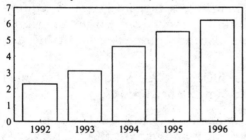

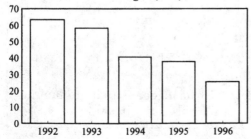

Figure 9-3. Examples of measurements for management review.

Employee Errors

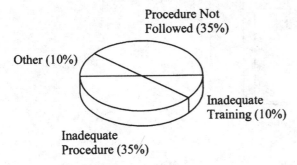

Procedure Not
Followed (35%)

Other (10%)

Inadequate
Training (10%)

Inadequate
Procedure (35%)

Equipment Failures

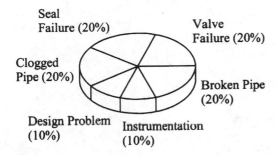

Seal
Failure (20%)

Valve
Failure (20%)

Clogged
Pipe (20%)

Broken Pipe
(20%)

Design Problem
(10%)

Instrumentation
(10%)

Figure 9-4. Nonconformance measurements.

walking around," or MBWA, is a popular management tool used at many facilities. By interacting with employees in their work area, upper management can observe environmental management practices first hand, and can seek employee suggestions about how to improve the EMS.

Another useful means of getting informal input is to engage in discussions with peer executives managing similar operations or issues. These types of discussions can help shape, reinforce, or reshape management approaches through the benefit of comparative analyses or benchmarking.

Finally, unscheduled reviews can take place as problems arise that need to be resolved immediately. Typically, some combination of communication via phone, e-mail, and person-to-person meetings will be

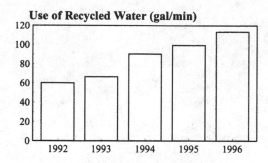

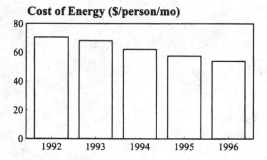

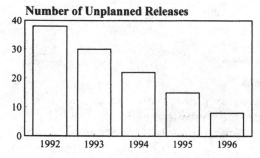

Figure 9-5. Additional examples of measurements for management review.

used to gather needed background, frame the problem or issue, and develop options for resolving and gaining agreement on an action plan. While resolving these types of incidents generally occurs outside the regular review meetings, the regular meetings are useful for tracking the status of corrective action items to closure and for ensuring that senior management is informed.

Management Review and
ISO 14001 Registration

While the environmental performance of an organization is not directly within the purview of ISO 14001 registrars, they can request evidence of environmental performance along with the relevant management reviews and follow-up actions. If environmental performance does not meet the EMS objectives and targets, that alone is not necessarily a problem if management is aware of the situation and is taking action to address the lack of progress. Modification of the objectives and targets is also a possible outcome, if there is technical or economic justification for doing so. If performance does not meet expectations, and management either is unaware or has not taken corrective action, then this could be construed as evidence that the EMS is not effective, and registration may be jeopardized.

10
Putting It
All Together

*ISO 14000 is significant for several reasons: It
provides us the means to globally manage our
environmental compliance according to an
internationally recognized norm; it will
provide efficiency and discipline and enable
operational integration with ISO 9000; it will
satisfy customer requirements for doing
business with environmentally responsible
suppliers; and it will provide evidence of due
diligence.*

JOEL CHARM
*Director, Occupational Health, Allied Signal,
Inc., and Chairman, U.S. Subtag on EMS*

The Plan-Do-Check-Act Process

The ISO 9000 quality management standards are built upon the plan-
do-check-act process. ISO 14001 also has this process embedded in its
major sections, including:

- Plan—Sections 4.1 and 4.2, Environmental Policy and Planning,
  respectively
- Do—Section 4.3, Implementation and Operation

- Check—Section 4.4, Checking and Corrective Action
- Act—Section 4.5, Management Review

This process allows for an integrated approach to achieving continual improvement of the environmental management system (EMS).

Establishing an Audit-Ready EMS Through Self-Assessment and Gap Analysis

The achievement of an audit-ready EMS requires that the documented environmental management program and all EMS elements and procedures be consistent with actual organization practices. Each requirement identified in ISO 14001 should be reviewed separately for adequate implementation. A tool for performing this self-assessment, or gap analysis, is presented in Appendix E. Although not required in ISO 14001, some organizations may want to create an environmental management manual. A sample manual is presented in Appendix F.

Developing a Registration Strategy

At some point, an organization must decide whether or not to pursue registration to ISO 14001. There are basically four options available to organizations that implement ISO 14001:

- Decide not to seek registration.
- Seek registration only after compelling reasons to do so exist.
- Obtain registration immediately.
- Self-declare conformance to ISO 14001.

Organizations electing the first option may or may not have an acceptable EMS in place or under development. These organizations, based on the information available to them, have decided that registration is not necessary for them to achieve marketplace objectives, environmental compliance objectives, and other environmentally related objectives.

Organizations that choose the second option want to implement an EMS for systematic environmental control, as well as a registration-ready posture, but do not necessarily want to spend money on third-party registration at this time (or at all, if it can be avoided). This middle-of-the-road position is currently shared by numerous U.S. businesses. These business organizations will use the next 6 to 18 months to ensure that the elements of their EMS are consistent with the requirements of ISO 14001. For many organizations with an existing, sophisticated EMS, the task may be an easy one. Other organizations will find the task both resource- and time-consuming. Nonetheless, implementing ISO 14001 requirements as soon as possible will allow an organization the flexibility to proceed with the registration process in the future, if it so desires. Ultimately, an organization that has laid the foundation for registration may be able to obtain it quickly to take advantage of new market opportunities that require it.

The third option is one organizations will pursue if they have a current or imminent market requirement to meet or if their organization otherwise perceives the benefits of registration as outweighing the costs. Market advantages may be derived from satisfying explicit requirements of U.S. or international customers. Registration can also be a catalyst for improving an organization's environmental management system, improving reliability of compliance processes, and identifying and reducing environmental risks.

The fourth option—that of self-declaring conformance to ISO 14001—will be selected by organizations that choose to proclaim their good-faith efforts for environmental protection, but get no specific benefit from registration. Examples of organizations that may choose this option include local governments, small service businesses, and nonprofit organizations.

A Roadmap to Registration

Once the organization decides to register to the ISO 14001 standard, a simple roadmap can be followed, as depicted in Fig. 10-1. Each of the elements in the roadmap is discussed in detail in other chapters or parts of this book, but it may be useful to visualize the basic requirements, as shown in their totality.

Registration

An EMS registration audit is needed in order to become registered to ISO 14001. The EMS registration audit will test for conformance of the organization's EMS to the requirements specified in the standard. Sample information that will be helpful for registration audits is presented in Table 10.1.

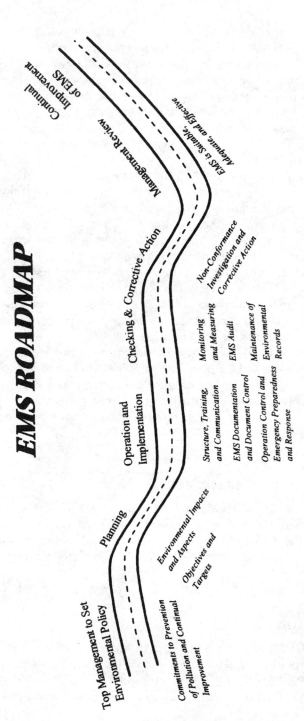

EMS ROADMAP

Top Management to Set Environmental Policy

Commitments to Prevention of Pollution and Continual Improvement

Planning

Environmental Impacts and Aspects

Objectives and Targets

Operation and Implementation

Structure, Training, and Communication

EMS Documentation and Document Control

Operation Control and Emergency Preparedness and Response

Checking & Corrective Action

Monitoring and Measuring

EMS Audit

Maintenance of Environmental Records

Non-Conformance Investigation and Corrective Action

Management Review

EMS is Suitable, Adequate, and Effective

Continual Improvement of EMS

Figure 10-1. Sample EMS roadmap.

Table 10-1. Examples of Information to Be Reviewed During an ISO 14001 Registration Audit

---

Environmental Policy
- Documented policy
- Policy development procedures (to ensure it comes from top management)
- Interviews with management and employees about existence of environmental policy
- Evidence about communication of environmental policy to public

Planning
- Procedures and/or matrix identifying environmental aspects, including updates and relevance compared with activities, products, and services
- Documents that specify legal requirements
- Documents specifying voluntary commitments
- Documents specifying industry standards
- Procedures, practices, and guidelines that enable compliance with country laws, voluntary commitments, and industry standards
- Documents or matrices that include objectives and targets, at each organizational level
- Documented activities related to achieving objectives and targets
- Internal reports, memos, meeting minutes, and other documentation related to planning, including information about objectives and targets, the environmental program, and other elements of environmental management
- Interviews with management and professionals involved with establishing objectives and targets
- Organization charts that detail responsibility for achieving objectives and targets
- Environmental program documentation including projects, resources, and plans, and other parts of the environmental program

Implementation and Operation
- Procedures, practices, matrices, and other planning documentation for allocating personnel, budgets, and other resources for environmental management
- Organization charts detailing roles, responsibilities, and authorities for environmental management
- Bulletin board announcements, internal news bulletins, on-line procedures, and other internal communications about roles, responsibilities, and authorities for environmental management
- Reports provided to top management about the performance of the environmental management system
- Training records, descriptions of employee job experience, and other documentation of employee awareness and competence
- Employee development plans
- Interviews with employees whose jobs have the potential to impact the environment, such as waste- and chemical-handling personnel, wastewater treatment plant operators, equipment specialists, maintenance technicians, and key manufacturing personnel, to ascertain level of environmental awareness and proficiency

(Continued)

Table 10-1. Examples of Information to Be Reviewed During an ISO 14001 Registration Audit (*Continued*)

- Observations of individuals in selected positions
- Procedures for communications between management and employees about environmental management issues
- Department meeting minutes, other meeting minutes, memos to employees, procedure sign-offs, and other documentation of communications between management and employees about environmental management issues
- Procedures for communications with external interested parties
- Newsletters, reports to shareholders, communications to neighborhood associations, and other evidence of communications to external interested parties
- Documentation of management's decisions regarding external communication
- Documents pertaining to the EMS
- Document control procedures and practices
- Records of document review and revision
- Flow charts, matrices, or other records that identify operations and activities
- Process hazard analyses, procedures, or other documentation identifying potential for accidents and emergency situations
- Emergency planning, response, and mitigation procedures
- HAZMAT and other emergency preparedness training records
- Records of incidents, emergency response actions, and corrective actions
- Records of emergency tests and drills
- Records of changes to emergency procedures, as necessary

Checking and Corrective Action

- Procedures for monitoring and measurement
- Records of monitoring and measurement
- Observations with respect to monitoring and measurement
- Meeting minutes, reports, memos, and other documents pertaining to the results of monitoring and measurement
- Procedures for compliance evaluation
- Reports and other documents pertaining to compliance evaluation
- Compliance evaluation schedule or matrix
- Procedures pertaining to EMS nonconformance and corrective action
- Self-assessment findings
- Employee communications about EMS nonconformance
- Documentation of corrective actions
- Records of changes to procedures as a result of findings of nonconformance
- Records of changes to operations as a result of findings of nonconformance
- List of environmental records incorporated into the EMS and documentation of storage practices
- Procedure for EMS audit program
- Audit schedules reports
- Audit reports
- Communications of audit results to management

Management Review

- Reports to management about the EMS

Table 10-1. Examples of Information to Be Reviewed During an ISO 14001 Registration Audit (*Continued*)

- Documents reviewed by management to ascertain the effectiveness and suitability of the EMS
- Records of management decisions with respect to the EMS after management review
- Documentation of changes to the EMS, as recommended by management
- Interviews with management about the management review process

Final Thoughts

The ISO 14001 specification provides a framework for establishing or improving an EMS. Beyond that, ISO 14001 provides an opportunity for organizations to integrate environmental management into the entire culture of the organization.

Some may view the requirements of ISO 14001 as a paperwork exercise that does not focus on the "right" things, such as environmental performance, technology, and regulatory compliance. The authors view ISO 14001 much differently. ISO 14001 creates a common international environmental language for global environmental progress. It leads to increased awareness among employees of their importance in the environmental management process. In the end, it is individual employees, more than their managers, who directly influence the environmental consequences of an organization's activities, products, and services. The ISO 14001 framework serves as a guide for willing and committed organizations that seek cultural change for better environmental care, more consistent and reliable compliance with laws and regulations, and better performance from their systems and operations.

List of Acronyms

ANSI: American National Standards Institute

API: American Petroleum Institute

ASTM: American Society for Testing and Materials

ASQC: American Society for Quality Control

BS: British Standard

BSI: British Standards Institute

CAG: Chairman's Advisory Group

CASCO: Conformity Assessment Committee

CD: committee draft

CEN: Committee for European Standardization

CENELEC: Committee for European Standardization for Electronics

CERTICO: Certification Committee of ISO

CMA: Chemical Manufacturers Association

CSA: Canadian Standards Association

DIN: German Institute for Standards

DIS: draft international standard

DFE: design for environment

DOE: Department of Energy

DOJ: Department of Justice

EA: environmental auditing

EAPS: environmental aspects in product standards

EAR: Environmental Auditing Roundtable

EDF: Environmental Defense Fund

EL: environmental labeling

EMAS: Eco-Management and Audit Scheme Regulation

EMS: environmental management system

EPA: Environmental Protection Agency

EPE: environmental performance evaluation

EU: European Union

GATT: General Agreement on Tariffs and Trade

IAF: International Accreditation Forum

IEC: International Electrotechnical Commission

ISO: International Organization for Standardization

LCA: life cycle assessment

NACCB: National Accreditation Council for Certifying Bodies

NAFTA: North American Free Trade Agreement

NEPA: National Environmental Policy Act

NIST: National Institute of Standards and Technology

OSHA: Occupational Safety and Health Administration

PPM: production and process methods

QSAR: Quality System Assessment Recognition (Group)

RAB: Registrar Accreditation Board

RBA: Registration Board for Assessors

SAGE: Strategic Advisory Group on the Environment

SARA: Superfund Amendments and Reauthorization Act

SC: subcommittee

SME: small to medium-size enterprise

SPI: Society of Plastics Industries

T&D: terms and definitions

TAG: Technical Advisory Group

TBT: Technical Barriers to Trade (Agreement)

TC: technical committee

TMB: Technical Management Board

TRI: Toxic Release Inventory

UKAS: United Kingdom Accreditation Service

UN: United Nations

UNCED: United Nations Conference on Environment and Development

WD: working draft

WG: working group

WICEMII: Second World Industry Conference on Environmental Management

WTO: World Trade Organization

USTR: U.S. Trade Representative

ISO Member Bodies and Designated Representative Organizations

ISO Member Bodies

Country	Designated Organization
Albania: DSC	Drejtoria e Stardardizimit dhe Cilesise Keshill I Ministrave
Algeria: INAPI	Institut Algérien de Normalisation et de Propriété Industriclle
Argentina: IRAM	Instituto Argentino de Racionalización de Materiales
Australia: SAA	Standards Australia
Austria: ON	Österreichisches Normungsinstiut
Bangladesh: BSTI	Bangladesh Standards and Testing Institution
Belarus: BELST	Committee for Standardization, Metrology, and Certification
Belgium: IBN	Institut Belge de Normalisation
Brazil: ABNT	Associação Brasileira de Normas Técnicas
Bulgaria: BDS	Committee for Standardization and Metrology at the Council of Ministers
Canada: SCC	Standards Council of Canada
Chile: INN	Instituto Nacional de Normalización
China: CSBTS	China State Bureau of Technical Supervision
Colombia: ICONTEC	Instituto Colombiano de Normas Técnicas
Croatia: DZNM	State Office for Standardization and Metrology
Cuba: NC	Comité Estatal de Normalización
Cyprus: CYS	Cyprus Organization for Standards and Control of Quality

Czech Republic: COSMT	Czech Office for Standards, Metrology, and Testing
Denmark: DS	Dansk Standard
Egypt: EOS	Egyptian Organization for Standardization and Quality Control
Ethiopia: ESA	Ethiopian Authority for Standardization
Finland: SFS	Finnish Standards Association
France: AFNOR	Association française de normalisation
Germany: DIN	Deutsches Institut für Normung
Greece: ELOT	Hellenic Organization for Standardization
Hungary: MSZH	Magyar Szabványügyi Hivatal
Iceland: STRI	Icelandic Council for Standardization
India: BIS	Bureau of Indian Standards
Indonesia: DSN	Dewan Standardisasi Nasional
Iran, Islamic Republic of: ISIRI	Institute of Standards and Industrial Research of Iran
Ireland: NSAI	National Standards Authority of Ireland
Israel: SII	Standards Institution of Israel
Italy: UNI	Ente Nazionale Italiano di Unificazione
Jamaica: JBS	Jamaica Bureau of Standards
Japan: JISC	Japanese Industrial Standards Committee
Kazakhstan: KAZMENST	Committee for Standardization, Metrology, and Certification
Kenya: KEBS	Kenya Bureau of Standards
Korea, Democratic People's Republic of: CSK	Committee for Standardization of the Democratic People's Republic of Korea
Korea, Republic of: KBS	Bureau of Standards
Libyan Arab Jamahiriya: LNCSM	Libyan National Centre for Standardization and Metrology
Malaysia: SIRIM	Standards and Industrial Research Institute of Malaysia
Mexico: DGN	Dirección General de Normas
Mongolia: MISM	Mongolian National Institute for Standardization and Metrology
Morocco: SNIMA	Service de normalisation industrielle marocaine
Netherlands: NNI	Nederlands Normalisatie-Institut
New Zealand: SNZ	Standards New Zealand
Norway: NSF	Norges Standardiseringsforbund
Pakistan: PSI	Pakistan Standards Institution
Philippines: BPS	Bureau of Product Standards
Poland: PKN	Polish Committee for Standardization
Portugal: IPQ	Instituto Português da Qualidade
Romania: IRS	Institutional Român de Standardizare
Russian Federation: GOST R	Committee of the Russian Federation for Standardization, Metrology, and Certification
Saudi Arabia: SASO	Saudi Arabian Standards Organization
Singapore: SISIR	Singapore Institute of Standards and Industrial Research
Slovakia: UNMS	Slovak Office of Standards, Metrology, and Testing
Slovenia: SMIS	Standards and Metrology Institute

South Africa: SABS	South African Bureau of Standards
Spain: AENOR	Asociación Española de Normalización y Certificatión
Sri Lanka: SLSI	Sri Lanka Standards Institution
Sweden: SIS	Standardiseringskommissio-nen I Sverige
Switzerland: SNV	Swiss Association for Standardization
Syrian Arab Republic: SASMO	Syrian Arab Organization for Standardization and Metrology
Tanzania, United Republic of: TBS	Tanzania Bureau of Standards
Thailand: TISI	Thai Industrial Standards Institute
Trinidad and Tobago: TTBS	Trinidad and Tobago Bureau of Standards
Tunisia: INNORPI	Institut national de la normalisation et de la propriété industrielle
Turkey: TSE	Türk Standardlari Enstitüsü
Ukraine: DSTU	State Committee of Ukraine for Standardization, Metrology, and Certification
United Kingdom: BSI	British Standards Institute
Uruguay: UNIT	Instituto Uruguayo de Normas Técnicas
USA: ANSI	American National Standards Institute
Uzbekistan: UZGOST	Uzbek State Centre for Standardization, Metrology, and Certification
Venezuela: COVENIN	Comisión Venezolana de Normas Industriales
Viet Nam: TCVN	Directorate for Standards and Quality
Yugoslavia: SZS	Savezni zavod za standardizaciju
Zimbabwe: SAZ	Standards Association of Zimbabwe

Correspondent Members

Country	Designated Organization
Bahrain	Directorate of Standards and Metrology
Barbados: BNSI	Barbados National Standards Institution
Brunei Darussalam	Construction Planning and Research Unit
Estonia: EVS	National Standards Board of Estonia
Hong Kong	Industry Department
Jordan: JDS	Directorate of Standards and Measures
Kuwait	Standards and Metrology Department
Lithuania: LST	Lithunian Standardization Office
Malawi: MBS Malawi	Malawi Bureau of Standards
Malta: MBS Malta	Malta Board of Standards
Mauritius: MSB	Mauritius Standards Bureau
Mozambique	National Institute of Standardization and Quality
Nepal	Nepal Bureau of Standards and Metrology
Oman	Directorate General for Specifications and Measurements
Papua New Guinea: PNGS	National Standards Council
Peru: INDECOPI	Instituto Nacional de Defensa de la Competencia y de la Protección de la Propiedad Intelectual
Qatar	Department of Standards and Measurements

Turkmenistan: MSIT	Major State Inspection of Turkmenistan
Uganda	Uganda National Bureau of Standards
United Arab Emirates: SSUAE	Directorate of Standardization and Metrology
Yemen: YSMO	Yemen Standardization, Metrology, and Quality Control Organization

Subscriber Members

Country	Designated Organization
Antigua and Barbuda: ABBS	Antigua and Barbuda Bureau of Standards
Bolivia: IBNORCA	Instituto Boliviano de Normalización y Calidad
Burundi: BBN	Bureau burundais de normalisation et contrôle de la qualité
Fiji	Fiji Trade Standards and Quality Control Office
Grenada: GDBS	Grenada Bureau of Standards
Saint Lucia	Saint Lucia Bureau of Standards

Examples of the United Nations Principles Pertaining to the Environment

United Nations Guiding Principles on the Human Environment—1972

Principle 1

Man has the fundamental right to freedom, equality, and adequate conditions of life, in an environment of a quality that permits a life of dignity and well-being, and he bears a solemn responsibility to protect and improve the environment for present and future generations. In this respect, policies promoting or perpetuating apartheid, racial segregation, discrimination, colonial and other forms of oppression and foreign domination stand condemned and must be eliminated.

Principle 2

The natural resources of the earth, including the air, water, land, flora, and fauna and especially representative samples of natural ecosystems, must be safeguarded for the benefit of present and future generations through careful planning or management, as appropriate.

Principle 3

The capacity of the earth to produce vital renewable resources must be maintained and, wherever practicable, restored or improved.

Principle 4

Man has a special responsibility to safeguard and wisely manage the heritage of wildlife and its habitat, which are now gravely imperilled by a combination of adverse factors. Nature conservation, including wildlife, must therefore receive importance in planning for economic development.

Principle 5

The nonrenewable resources of the earth must be employed in such a way as to guard against the danger of their future exhaustion and to ensure that benefits from such employment are shared by all mankind.

Principle 6

The discharge of toxic substances or of other substances and the release of heat, in such quantities or concentrations as to exceed the capacity of the environment to render them harmless, must be halted in order to ensure that serious or irreversible damage is not inflicted upon ecosystems. The just struggle of the peoples of all countries against pollution should be supported.

Principle 7

States shall take all possible steps to prevent pollution of the seas by substances that are liable to create hazards to human health, to harm living resources and marine life, to damage amenities or to interfere with other legitimate uses of the sea.

Principle 8

Economic and social development is essential for ensuring a favorable living and working environment for man and for creating conditions on earth that are necessary for the improvement of the quality of life.

Principle 9

Environmental deficiencies generated by the conditions of underdevelopment and natural disasters pose grave problems and can best be remedied by accelerated development through the transfer of substantial quantities of financial and technological assistance.

Principle 10

For the developing countries, stability of prices and adequate earning for primary commodities and raw materials are essential to environmental management since economic factors as well as ecological processes must be taken into account.

United Nations Principles on Environment and Development—1992

Principle 1

Human beings are at the center of concerns for sustainable development. They are entitled to a healthy and productive life in harmony with nature.

Principle 2

States have, in accordance with the Charter of the United Nations and the principles of international law, the sovereign right to exploit their own resources pursuant to their own environmental and developmental policies, and the responsibility to ensure that activities within their jurisdiction or control do not cause damage to the environment of other States or of areas beyond the limits of national jurisdiction.

Principle 3

The right to development must be fulfilled so as to equitably meet developmental and environmental needs of present and future generations.

Principle 4

In order to achieve sustainable development, environmental protection shall constitute an integral part of the development process and cannot be considered in isolation from it.

Principle 5

All States and all people shall cooperate in the essential task of eradicating poverty as an indispensable requirement for sustainable development, in order to decrease the disparities in standards of living and better meet the needs of the majority of the people of the world.

Principle 6

The special situation and needs of developing countries, particularly the least developed and those most environmentally vulnerable, shall be given special priority. International actions in the field of environment and development should also address the interest and needs of all countries.

Principle 7

States shall cooperate in a spirit of global partnership to conserve, protect and restore the health and integrity of the Earth's ecosystem. In view of the different contributions to global environmental degradation, States have common but differentiated responsibilities. The developed countries acknowledge the responsibility that they bear in the international pursuit of sustainable development in view of the pressures their societies place on the global environment and of the technologies and financial resources they command.

Principle 8

To achieve sustainable development and a high quality of life for all people, States should reduce and eliminate unsustainable patterns of production and consumption and promote appropriate demographic policies.

Principle 9

States should cooperate to strengthen endogenous capacity-building for sustainable development by improving scientific understanding through exchanges of scientific and technological knowledge, and by enhancing the development, adaptation, diffusion and transfer of technologies, including new and innovative technologies.

Principle 10

Environmental issues are best handled with the participation of all concerned citizens, at the relevant level. At the national level, each individual shall have appropriate access to information concerning the environment that is held by public authorities, including information on hazardous materials and activities in their communities, and the opportunity to participate in decision-making processes. States shall facilitate and encourage public awareness and participation by making information widely available. Effective access to judicial and administrative proceedings, including redress and remedy, shall be provided.

International Chamber of Commerce Principles for Environmental Management

1. Corporate priority

To recognise environmental management as among the highest corporate priorities and as a key determinant to sustainable development; to establish policies, programmes and practices for conducting operations in an environmentally sound manner.

2. Integrated management

To integrate these policies, programmes and practices fully into each business as an essential element of management in all its functions.

3. Process of improvement

To continue to improve corporate policies, programmes and environmental performance, taking into account technical developments, scientific understanding, consumer needs and community expectations, with legal regulations as a starting point; and to apply the same environmental criteria internationally.

4. Employee education

To educate, train and motivate employees to conduct their activities in an environmentally responsible manner.

5. Prior assessment

To assess environmental impacts before starting a new activity or project and before decommissioning a facility or leaving a site.

6. Products and services

To develop and provide products or services that have no undue environmental impact and are safe in their intended use, that are efficient in their consumption of energy and natural resources and that can be recycled, reused or disposed of safely.

7. Customer advice

To advise, and where relevant educate, customers, distributors and the public in the safe use, transportation, storage and disposal of products provided; and to apply similar considerations to the provision of services.

8. Facilities and operations

To develop, design and operate facilities and conduct activities taking into consideration the efficient use of energy and materials, the sustainable use of renewable resources, the minimisation of adverse environmental impact and waste generation, and the safe and responsible disposal of residual wastes.

9. Research

To conduct or support research on the environmental impacts of raw materials, products, processes, emissions and wastes associated with the enterprise and on the means of minimizing such adverse impacts.

10. Precautionary approach

To modify the manufacture, marketing or use of products or services or the conduct of activities, consistent with scientific and technical understanding, to prevent serious or irreversible environmental degradation.

11. Contractors and suppliers

To promote the adoption of these principles by contractors acting on behalf of the enterprise, encouraging and, where appropriate, requiring improvements in their practices to make them consistent with those of the enterprise; and to encourage the wider adoption of these principles by suppliers.

12. Emergency preparedness

To develop and maintain, where significant hazards exist, emergency preparedness plans in conjunction with the emergency services, relevant authorities and the local community, recognizing potential transboundary impacts.

13. Transfer of technology

To contribute to the transfer of environmentally sound technology and management methods through the industrial and public sectors.

14. Contributing to the common effort

To contribute to the development of public policy and to business, governmental and intergovernmental programmes and educational initiatives that will enhance environmental awareness and protection.

15. Openness to concerns

To foster openness and dialogue with employees and the public, anticipating and responding to their concerns about the potential hazards and impacts of operations, products, wastes or services, including those of transboundary or global significance.

16. Compliance and reporting

To measure environmental performance; to conduct regular environmental audits and assessments of compliance with company requirements, legal requirements and these principles; and periodically to provide appropriate information to the Board of Directors, shareholders, employees, the authorities and the public.

ISO 14001 Self-Assessment and Gap Analysis

Section 4.0   General

The Organization has established an environmental management system (EMS) that meets the requirements of the standard.

☐ Fully established and implemented

☐ Established and some requirements implemented

☐ Not established or implemented

Section 4.1   Environmental Policy

Top management has defined the organization's environmental policy.

☐ Yes

☐ No

The environmental policy is appropriate and considers the nature, scale and environmental impacts of the organization's activities, products and services.

☐ Fully appropriate

☐ Partially appropriate, with some aspects/impacts omitted

☐ Not appropriate

The environmental policy includes a commitment to continual improvement.

☐ Yes, fully

☐ Commitment to continual improvement in policy could be improved

☐ No commitment to continual improvement

The environmental policy includes a commitment to prevention of pollution.

☐ Yes, fully

☐ Commitment to prevention of pollution in policy could be improved

☐ No commitment to prevention of pollution

The environmental policy includes a commitment to comply with applicable environmental legislation and regulations.

☐ Yes

☐ Yes, but commitment could be improved

☐ No

The environmental policy includes a commitment to comply with other requirements to which the organization subscribes.

☐ Yes

☐ Yes, but commitment could be improved

☐ No

The environmental policy provides a framework for setting and reviewing environmental objectives and targets.

☐ Yes

☐ Yes, but framework could be improved

☐ No

The environmental policy is documented and implemented.

☐ Both

☐ The policy is documented, but not fully implemented

☐ Neither

The environmental policy is maintained and communicated to all employees.

☐ Both

☐ Maintained, but communication efforts could be improved

☐ Neither

The environmental policy is available to the public.

☐ Yes

☐ No

Section 4.2   Planning

Environmental Aspects

There is a procedure established and maintained to identify the organization's environmental aspects in order to determine which aspects have significant impacts on the environment.

☐ Procedure exists

☐ Procedure could be improved

☐ Procedure does not exist

Significant aspects are considered when setting environmental objectives.

☐ Yes, all

☐ Some

☐ No, none

Information pertaining to significant aspects is kept up to date.

☐ Yes

☐ No

Legal and Other Requirements

A procedure has been established to identify and have access to legal requirements and other requirements to which the organization subscribes that are directly applicable to environmental impacts. The procedure is current and is maintained.

- ☐ Yes, a procedure exists and is current
- ☐ A procedure exists, but need to be updated
- ☐ No, a procedure does not exist

Objectives and Targets

Objectives and targets have been established at each relevant function and level within the organization.

- ☐ Yes, fully
- ☐ Partially, at some functions and levels, but not all
- ☐ No

Relevant legal and other requirements were considered when establishing objectives and targets.

- ☐ Yes
- ☐ No

Significant environmental aspects were considered when establishing objectives and targets.

- ☐ Yes, fully
- ☐ Partially
- ☐ No

Technological options, financial, operational and business requirements were considered when establishing objective and targets.

- ☐ Yes, fully
- ☐ Partially
- ☐ No

Views of interested parties were considered when establishing objectives and targets.

- ☐ Yes, fully
- ☐ Partially

☐ No

The objectives and targets are consistent with the environmental policy.

☐ Yes, fully
☐ Partially
☐ No

The objectives and targets are consistent with the commitment to prevention of pollution.

☐ Yes, fully
☐ Partially
☐ No

Environmental Management Programs

There is an established environmental management program for achieving environmental objectives and targets.

☐ Yes, fully established
☐ Partially established
☐ No

The environmental management program includes a designation of responsibility for achieving objectives and targets at each relevant function and level of the organization.

☐ Yes
☐ Some responsibilities not designated
☐ No

The environmental management program includes the means and time frame by which the objectives and targets are to be achieved.

☐ Yes
☐ Some time frames not included
☐ No

The environmental management program applies to new developments, new or modified activities, products and services, as appropriate.

☐ Yes, fully
☐ Partially
☐ No

Section 4.3
Implementation and
Operation

Structure and Responsibility

Roles, responsibility, and authorities are defined, documented and communicated.

☐ Yes, fully
☐ Partially
☐ No

Resources essential to the implementation and control of the EMS are provided—including human resources and specialized skills, technology and financial resources.

☐ Yes, fully
☐ Partially
☐ No

Top management has appointed a specific management representative(s) with defined roles, responsibilities and authority for establishing, implementing and maintaining the EMS.

☐ Yes
☐ Some representatives not appointed
☐ Some roles, responsibilities and authorities not defined
☐ No

These representatives report to top management the performance of the environmental management system for management review and as a basis for continual improvement.

☐ Yes, on a scheduled basis

☐ Sometimes, but not on a scheduled basis

☐ No

Training, Awareness and Competence

Training needs have been identified and appropriate personnel have received necessary training.

☐ Yes, fully

☐ Partially

☐ No

Procedures are established and maintained to make employees aware of the importance of conformance with the environmental policy and procedures and with the requirements of the EMS.

☐ Yes, procedures are established and maintained

☐ Procedures could be improved

☐ No

Procedures are established and maintained to make employees aware of significant impacts, actual or potential, of their work activities and the environmental benefits of improved personal performance.

☐ Yes, procedures are established and maintained

☐ Procedures could be improved

☐ No

Procedures are established and maintained to make employees aware of their roles and responsibilities in achieving conformance with the environmental policy and with the requirements of the EMS—including emergency preparedness and response requirements.

☐ Yes, procedures are established and maintained

☐ Procedures could be improved

☐ No

Procedures are established and maintained to make employees aware of the potential consequences of nonadherence to operating procedures.

☐ Yes, procedures are established and maintained
☐ Procedures could be improved
☐ No

Personnel who perform tasks that may cause significant environmental impacts are competent to perform their duties based on education, training or experience.

☐ Yes, fully competent
☐ Partially competent
☐ No

Communication

Procedures are established and maintained for internal communication about significant environmental aspects and the EMS.

☐ Yes, procedures are established and maintained
☐ Procedures could be improved
☐ No

Procedures are established and maintained for receiving, documenting and responding to relevant communication from external interested parties as it relates to significant environmental aspects and the EMS.

☐ Yes, procedures are established and maintained
☐ Procedures could be improved
☐ No

Means for externally communicating information about significant environmental aspects have been considered and documented.

☐ Yes, fully reviewed and documented
☐ Reviewed, but not documented
☐ No

EMS Documentation

Information describing the core elements of the EMS is established and maintained.

☐ Yes, fully

☐ Partially

☐ No

Information that provides direction to related documentation is established and maintained.

☐ Yes, fully

☐ Partially

☐ No

Document Control

Procedures for controlling all documents are established, maintained and readily available.

☐ Yes, procedures are established, maintained and readily available

☐ Procedures established, but not readily available

☐ Procedures could be improved

☐ Procedures not established

These procedures are periodically reviewed, revised if necessary, and approved by authorized personnel.

☐ Yes, procedures are periodically reviewed and revised

☐ Procedures are reviewed and revised, but not on any specific schedule

☐ No, procedures are not reviewed

Current versions of relevant documents are available and in their proper locations for effective functioning of the EMS.

☐ Yes, fully

☐ Partially

☐ No

Obsolete documents are promptly removed from all areas using these documents.

☐ Yes
☐ No

Obsolete documents retained for legal or knowledge preservation purposes are so marked.

☐ Yes
☐ No

Documents are legible, dated and readily identifiable.

☐ Yes
☐ No

There are procedures and responsibilities established and maintained for creating and modifying pertinent documents.

☐ Yes, procedures are established and maintained
☐ Procedures could be improved
☐ No

Operational Control

Operations and activities that are associated with significant environmental impacts and which fall within the scope of the environmental policy, objectives and targets have been identified.

☐ Yes, fully
☐ Partially
☐ No

Procedures pertaining to these activities are established and maintained to cover situations that, in their absence, might lead to deviations from the environmental policy and the objectives and targets.

☐ Yes, procedures are established and maintained
☐ Procedures could be improved
☐ No

Procedures stipulate operating criteria.

- ☐ Yes
- ☐ No

Procedures related to the significant environmental aspects of good and services from suppliers and contractors are established and maintained.

- ☐ Yes, procedures are established and maintained
- ☐ Procedures could be improved
- ☐ No

Relevant procedures and requirements are communicated to suppliers and contractors.

- ☐ Yes, fully
- ☐ Partially
- ☐ No

Emergency Preparedness and Response

Procedures that identify the potential for and the response to accidents and emergency situations are established and maintained.

- ☐ Yes, procedures are established and maintained
- ☐ Procedures could be improved
- ☐ No

The procedures address prevention and mitigation of environmental impacts that may be associated with any accidents or emergency situations.

- ☐ Yes, fully
- ☐ Partially
- ☐ No

Emergency preparedness and response procedures are reviewed and revised as necessary, but in particular after the occurrence of accidents or emergency situations.

- ☐ Yes
- ☐ No

Emergency preparedness and response procedures are periodically tested where practicable.

☐ Yes
☐ No

Section 4.4   Checking and Corrective Action

Monitoring and Measurement

There are procedures established and maintained to monitor and measure on a regular basis the key characteristics of the operations and activities that can have a significant impact on the environment.

☐ Yes, procedures are established and maintained
☐ Procedures could be improved
☐ No

Monitoring and measurement includes recording information to track performance, relevant operations controls and conformance with objectives and targets.

☐ Yes, fully
☐ Partially
☐ No

Monitoring equipment is calibrated and maintained and a record of the calibration process is retained, per procedure.

☐ Yes
☐ No

A procedure is established and maintained to periodically evaluate compliance with relevant environmental legislation and regulations.

☐ Yes, a procedure is established and maintained
☐ Procedure could be improved
☐ No

Nonconformance and Corrective and Preventive Action

Procedures are established and maintained for handling and investigating nonconformance, for taking action to mitigate the impacts caused by nonconformance, and for initiating corrective and preventive action.

☐ Yes, procedures are established and maintained
☐ Procedures could be improved
☐ No

Responsibility and authority for these same tasks are defined.

☐ Yes, fully
☐ Partially
☐ No

Any corrective or preventive action is appropriate for the magnitude of actual or potential environmental impact that has or could occur from nonconformance.

☐ Yes
☐ No

Procedures are modified to reflect corrective and preventive action.

☐ Yes, fully
☐ Partially
☐ No

Records

Procedures are established and maintained for the identification, maintenance and disposition of environmental records.

☐ Yes, procedures are established and maintained
☐ Procedures could be improved
☐ No

Environmental records include training records, records of audit results and records of management reviews.

☐ Yes
☐ No

Environmental records are legible, identifiable and traceable to the activity, product or service involved.

☐ Yes
☐ No

Environmental records are easily retrievable and are protected from damage, deterioration or loss.

☐ Yes
☐ No

The retention history of the records is documented.

☐ Yes
☐ No

The records demonstrate compliance with the standard.

☐ Yes, fully
☐ Partially
☐ No

EMS Audit

A program and procedures are established and maintained for periodic EMS audits.

☐ Yes, a program and procedures are established and maintained
☐ The program and procedures could be improved
☐ A program and procedures have not been established

The audits determine whether or not the EMS conforms to specified internal requirements for environmental management, including conformance to the requirements of the standard.

☐ Yes, fully

☐ Partially

☐ No

The audits determine whether or not the EMS has been properly implemented and maintained.

☐ Yes, fully

☐ Partially

☐ No

The audit results are presented to management for review.

☐ Yes

☐ No

The audit procedures cover the audit scope, frequency and methodologies, and responsibilities and requirements for conducting audits and reporting results.

☐ Yes, fully

☐ Partially

☐ No

Section 4.5 Management Review

Top management periodically reviews the EMS to ensure continuing suitability, adequacy and effectiveness.

☐ Yes, on a scheduled basis

☐ Sometimes, but not on a regular basis

☐ No

Necessary information is collected and provided to allow management to carry out the evaluation.

☐ Yes, fully
☐ Partially
☐ No

Management assesses the need for changes in environmental policy, objectives, and in the EMS, as indicated by EMS audit results, changing circumstances, and the commitment to continual improvement.

☐ Yes, fully
☐ Partially
☐ No

Appendix F
Sample Environmental Management Manual

Table of Contents

0.0 General

0.1 Company Description

0.2 Manual Control and Record of Revision

0.3 Manual Distribution

1.0 Environmental Policy

2.0 Planning

2.1 Environmental Aspects

2.2 Legal and Other Requirements

2.3 Objectives and Targets

2.4 Environmental Protection Program

3.0 Implementation and Operation

3.1 Structure and Responsibility

3.2 Training

3.3 Communication

3.4 Environmental System Documentation

3.5 Document Control

3.6 Operational Control

3.7 Emergency Preparedness and Response

4.0 Checking and Corrective Action

4.1 Monitoring and Measurement

4.2 Nonconformance and Corrective and Preventive Action

4.3 Records

4.4 Environmental Management System Audit

5.0 Management Review

Prepared by: John Engineer	Issue Number: 3
Approved by: Jane Manager	Issue Date: 09/15/95

0.0 General

0.1 Company Description

New Process, Inc. manufactures a variety of components for other companies which then assemble them into their own products. For example, New Process may supply a computer chassis assembly ready for the client to fill with computer parts. New Process subcontracts much of its work to keep fixed costs low and to increase responsiveness to customer's needs.

New Process' clients include major electronics and instrumentation firms located around the world. Headquartered in San Jose, California, New Process was founded in 1985. Its divisions include X, Y, and Z Products.

New Process is committed to production, product, and environmental quality. The company is ISO 9000 certified and has applied for the Malcolm Baldrige Quality Award. Total quality management and client service are fundamental elements of the company's culture. This environmental management manual incorporates and reflects the environmental protection ethic that is fostered as part of the company's culture and heritage.

0.2 Manual Control and Record of Revision

This manual is maintained by the company's Office of Environmental Protection, which is currently headed by a Corporate Vice President. The originals of all pages of this manual shall be kept by this office, which will be the final authority on the manual content. Any revisions to this manual shall be routed to, authorized by, and distributed from this office.

Prepared by: John Engineer	Issue Number: 3
Approved by: Jane Manager	Issue Date: 09/15/95

Record of Revisions

Date	Section	Page	Summary of Change	Signature

0.3 Manual Distribution

The below-listed offices and functions will be issued duplicates of this manual, which they will maintain current with periodic updates received from the Corporate Office of Environmental Protection:

1. Company President
2. Company Senior Vice President
3. Company Vice President of Finance
4. Heads of Manufacturing Divisions X, Y, and Z
5. Heads of Facility Operations and Engineering for Divisions X, Y, and Z
6. Director of Corporate Purchasing
7. Division Heads of Environmental Protection
8. Managers of Division Purchasing

Prepared by: John Engineer	Issue Number: 3
Approved by: Jane Manager	Issue Date: 09/15/95

1.0 Environmental Policy

New Process, Inc., strives to achieve "low environmental impact" manufacturing. It aims to continuously find new ways to reduce quantities of materials, emissions, and energy required to produce products. New Process is proud of its achievements, which demonstrate a firm and steady commitment to excellence since its founding.

New Process' environmental policy calls for:

- Products that are designed and manufactured safely for employees, clients, contractors, neighbors, and end users

- Manufacturing techniques that maximize the use of benign processes while minimizing emissions associated with chemical use. Prevention of pollution guides all design and manufacturing decisions and is the preferred option.

- Continual improvements in operations, emergency preparedness and management oversight to increase the effectiveness and reliability of the management system

- Identifying relevant safety and environmental requirements that are consistently met

- Progress at meeting environmental goals, objectives and targets established through internal and external assessments and reviews

New Process makes this policy available and accessible to all its employees and publishes it externally for the public.

Prepared by: John Engineer	Issue Number: 2
Approved by: Jane Manager	Issue Date: 05/8/95

2.0 Planning

2.1 Environmental Aspects

2.1.1 Scope. The process described in this section conforms to ISO 14001, Section 4.2.1, Environmental Aspects. This process is followed to identify the environmental aspects of the activities, products, and services of New Process, Inc.

2.1.2 Responsibility. Each operating division is required to identify the environmental aspects of its activities, products, and services and determine which have or can have significant environmental impacts. The Environmental Protection function will provide preparation assistance, guidance, and review of results.

2.1.3 Process. Determination of significant environmental aspects is done prior to:

- Setting objectives and targets
- The start-up of a new product or process
- Modification of an existing product or process which creates new environmental aspects or significantly increases existing environmental impacts

Updates shall be made as they occur. A comprehensive update shall be prepared at least every 3 years.

2.1.4 Related Documentation. Operating procedures include the most recent version of:

OP 4.2.1 Product and Process Environmental Aspects and Impact Assessment

Prepared by: John Engineer	Issue Number: 2
Approved by: Jane Manager	Issue Date: 09/23/95

2.2 Legal and Other Requirements

2.2.1 Scope. The process described in this section conforms to ISO 14001, Section 4.2.2, Legal and Other Requirements. This process is followed to identify the legal and other requirements directly applicable to significant environmental aspects listed in Section 2 above.

2.2.2 Responsibility. Each operating division is required to identify the legal and other environmental requirements which are directly applicable to its operations. The Divisional Environmental Protection function will provide assistance and periodically review the compilation to ensure its accuracy.

2.2.3 Process. Determination of applicable legal and other environmental requirements shall be completed for significant environmental aspects when:

- This has not previously been done
- A new requirement is promulgated or an existing requirement changes
- A new product, process, or service is planned
- An existing product or process is to be modified

A comprehensive calendar of compliance obligations shall be prepared and kept current. The status and satisfaction of these obligations shall be periodically reviewed with management.

2.2.4 Related Documentation. Operating procedures include the most recent version of:

OP 4.2.1 Product and Process Environmental Aspects and Impact Assessment

OP 4.2.4 Environmental Protection Program Planning

Prepared by: John Engineer	Issue Number: 2
Approved by: Jane Manager	Issue Date: 09/23/95

2.3 Objectives and Targets

2.3.1 Scope. The process described in this section conforms to ISO 14001, Section 4.2.3, Objectives and Targets. This process is followed to establish the environmental objectives and targets for New Process to attain the goals of its environmental policy.

2.3.2 Responsibility. Each operating division is required to set objectives and targets to attain the goals of New Process' environmental policy. The Divisional Environmental Protection function will provide guidance in setting, and periodically reviewing, these environmental objectives and targets.

2.3.3 Process. Establishing relevant objectives and targets for attaining the goals of our environmental policy begins with the commitment of senior management within each division. Environmental objectives and targets shall be included in the division's strategic plan.

2.3.4 Related Documentation. Operating procedures include the most recent version of:

OP 4.2.4 Environmental Protection Program Planning

2.4 Environmental Protection Program

2.4.1 Scope. The process described in this section conforms to ISO 14001, Section 4.2.4, Environmental Management Programme. This process is followed to structure the environmental protection program used by New Process, Inc. to attain the goals of its environmental policy.

2.4.2 Responsibility. Each operating division is required to establish an environmental protection program consistent with this manual which will guide their operations to attain the goals of New Process'

Prepared by: John Engineer	Issue Number: 2
Approved by: Jane Manager	Issue Date: 09/23/95

environmental policy. The Divisional Environmental Protection function will provide guidance in creating, and periodically reviewing, division-specific environmental protection programs.

2.4.3 Process. Creating an environmental protection program involves:

- Setting and articulating a vision of operational integrity and thoroughness
- Identifying program elements applicable to the division's operations
- Providing adequate human and financial resources to effectively execute the program
- Incorporating continuous improvement techniques, including regular measurement and monitoring reports

2.4.4 Related Documentation. Operating procedures include the most recent version of:

Environmental Management Manual [this document]

OP 4.2.4 Environmental Protection Program Planning

3.0 Implementation and Operation

3.1 Structure and Responsibility

3.1.1 Scope. The process described in this section conforms to ISO 14001, Section 4.3.1, Structure and Responsibility. This process establishes roles, responsibilities, and authorities to allow New Process, Inc. to meet the goals established in its environmental policy and in the associated objectives and targets.

3.1.2 Responsibility. Top management has established the overall company organization, including roles, responsibilities, and authorities for environmental management. The Vice President of Environmental Protection is designated as the company focal point responsible for environmental management, including:

- Ensuring that environmental management system requirements are established, implemented, and maintained in accordance with this standard
- Reporting on the performance of the environmental management system to top management for review and to improve the environmental management system

Each operating division is required to establish the roles and responsibilities within its organizations which will guide their operations to attain the goals of New Process' environmental policy. Each operating division shall apply sufficient resources (human, technological, and financial) to effectively implement its environmental management system.

3.1.3 Process. Division and corporate management are required to establish roles, responsibilities, and authorities for environmental

Prepared by: John Engineer	Issue Number: 2
Approved by: Jane Manager	Issue Date: 09/23/95

management integrated with the rest of the organization. Corporate management is required to identify those environmental issues of potential strategic advantage to New Process.

3.1.4 Related Documentation. Operating procedures include the most recent version of:

Environmental Management Manual [this document]

OP 4.2.4 Environmental Protection Program Planning

3.2 Training

3.2.1 Scope. The process described in this section conforms to ISO 14001, Section 4.3.2, Training, Awareness, and Competence. This process is followed to identify training needs and for the delivery of training to all appropriate personnel. Employees at all relevant levels shall be made aware of:

- The importance of conformance with the environmental policy and procedures and with the requirements of the environmental management system
- The significant environmental impacts, actual or potential, of their work activities and the environmental benefits of improved personal performance
- Their roles and responsibilities in achieving conformance with the environmental policy and procedures and with the requirements of the environmental management system, including emergency preparedness and response requirements
- The potential consequences of departure from specific operating procedures

Personnel performing tasks which can cause significant environmental impacts are required to be evaluated for competence on the basis of appropriate education, training, and/or experience.

| Prepared by: John Engineer | Issue Number: 2 |
| Approved by: Jane Manager | Issue Date: 09/23/95 |

3.2.2 Responsibility. The Corporate Environmental Protection function establishes the framework for a training program which meets the above requirements. Division management is responsible for providing needed training of employees, contractors, and suppliers.

3.2.3 Process. The Corporate Environmental Protection function will provide the training framework (contained in OP 4.3.2) to the operating divisions. Each division will arrange for training of employees, contractors, and suppliers. Divisions are required to keep training records.

3.2.4 Related Documentation. Operating procedures include the most recent version of:

OP 4.3.2 Environmental Training

3.3 Communication

3.3.1 Scope. The process described in this section conforms to ISO 14001, Section 4.3.3, Communication.

This process allows for:

- Internal communication between various functions and levels of the organization
- Receiving, documenting, and responding to relevant communication from external interested parties
- External communication regarding New Process' significant environmental aspects

3.3.2 Responsibility. The Corporate Communications function, in conjunction with the Corporate Environmental Protection function, is required to define a process to effectuate the communications requirements of the EMS. That function will oversee its implementation and upkeep.

Prepared by: John Engineer	Issue Number: 2
Approved by: Jane Manager	Issue Date: 09/23/95

3.3.3 Process. Operating divisions are required to implement programs to fulfill the communications requirements of the EMS. Close liaison and coordination will be maintained with both the Corporate Communications Environmental Protection functions.

3.3.4 Related Documentation. Operating procedures include the most recent version of:

OP 4.3.3 Environmental Communication

3.4 Environmental System Documentation

3.4.1 Scope. The process described in this section conforms to ISO 14001, Section 4.3.4, Environmental System Documentation. This process is followed to establish the framework to manage information, in paper or electronic form, to:

- Describe the elements of the information management system and their interaction
- Manage documentation related to the EMS

3.4.2 Responsibility. The Corporate Environmental Protection function is required to define this process with input and support from appropriate groups, including Information Solutions, Administrative Support, and Legal. Information Solutions is responsible for maintaining the information management system.

3.4.3 Process. The Corporate Environmental Protection function is required to identify information management elements, such as regulatory data and reporting mandates and the information requirements of the EMS. Operating Procedure 4.3.4 contains details of the current New Process information management process and the associated information system. A companywide system is used to foster operating

Prepared by: John Engineer	Issue Number: 2
Approved by: Jane Manager	Issue Date: 09/23/95

flexibility and efficiency. In order to accommodate necessary changes in information requirements and information system capabilities, this operating procedure and associated system are designed to be adaptable and flexible.

3.4.4 Related Documentation. Operating procedures include the most recent version of:

OP 4.3.4 Environmental System Information Management

3.5 Document Control

3.5.1 Scope. The process described in this section conforms to ISO 14001, Section 4.3.5, Document Control. This process is followed to establish the framework to control electronic and paper documents so that:

- They can be located
- They are periodically reviewed, revised as necessary, and approved for adequacy by authorized personnel
- The current versions of relevant documents are available at all locations where operations essential to the effective functioning of the system are performed
- Obsolete documents are promptly removed from all points of issue and points of use or otherwise treated to prevent unwanted use
- Any obsolete documents retained for legal and/or historical purposes are suitably identified

3.5.2 Responsibility. The Corporate Environmental Protection function is required to define this process, with input and support from appropriate groups, including Information Solutions, Administrative Support, and Legal. Divisions are required to implement processes to fulfill the document control aspects of the EMS.

Prepared by: John Engineer	Issue Number: 2
Approved by: Jane Manager	Issue Date: 09/23/95

3.5.3 Process. The fundamental strategy of New Process, Inc. is to utilize the information management system to store and retrieve current information electronically. Paper documentation originating from this electronic storage and retrieval system shall be considered potentially out of date, unless verified by the user as current at each time of use.

3.5.4 Related Documentation. Operating procedures include the most recent version of:

> OP 4.3.4 Environmental System Information Management (Includes Document and Records Control)

3.6 Operational Control

3.6.1 Scope. The process described in this section conforms to ISO 14001, Section 4.3.6, Operational Control. This process is followed to establish a framework to control operations so that:

- New Process' environmental policy is met or surpassed
- New Process' environmental objectives and targets are met or surpassed, where feasible. Procedures are established and maintained to:

> Identify activities, products, and services which may have significant environmental impacts
> Address situations in which deviations from the environmental policy, targets, or objectives occur
> Communicate relevant requirements to suppliers and contractors

3.6.2 Responsibility. The Corporate Environmental Protection function will define the framework to fulfill the operational control elements of the EMS. The Divisional Environmental Protection function will give assistance to help its division develop adequate operating procedures. Each operating division is required to identify its activities, products, and services related to significant environmental

Prepared by: John Engineer	Issue Number: 2
Approved by: Jane Manager	Issue Date: 09/23/95

impacts and to develop and maintain operating procedures to address those aspects.

3.6.3 Process. Each division identifies its activities, products, and services related to significant environmental impacts, and implements operating procedures to protect the environment and meet or surpass the company's environmental policy. The Divisional Environmental Protection function will, at periodic intervals, check the operating procedures to ensure they are complete and effective.

3.6.4 Related Documentation. Operating procedures include the most recent version of:

OP 4.2.1 Product and Process Environmental Aspects and Impact Assessment

OP 4.3.6 Environmental Evaluation of Suppliers

Operating Procedures

3.7 Emergency Preparedness and Response

3.7.1 Scope. The process described in this section conforms to ISO 14001, Section 4.3.7, Emergency Preparedness and Response. Prevention of incidents and accidents is addressed in other sections in this manual. This process is followed to establish the framework to prepare for and respond to emergency situations which may create a threat to health, safety, or the environment.

3.7.2 Responsibility. The Corporate Environmental Protection function is required to provide guidelines to each operating division for use in identification of operations and activities that could cause significant environmental impact should an incident occur. The Divisional Environmental Protection function will assist its division in assessing its operations. Each operating division will identify its oper-

Prepared by: John Engineer	Issue Number: 3
Approved by: Jane Manager	Issue Date: 10/25/95

ations and activities which may cause significant environmental impact should an incident occur. Additionally, each operating division will develop and maintain operating procedures which address preparedness, prevention, and response to incidents that create a threat to health, safety, or the environment.

3.7.3 Process. Each operating division identifies its operations which may cause significant environmental impact should an incident occur. The division prepares procedures that incorporate adequate measures for emergency preparedness, prevention, and response to incidents. The Divisional Environmental Protection function will, at periodic intervals, check the operating division's procedures to ensure they are complete and effective.

3.7.4 Related Documentation. Operating procedures include the most recent version of:

OP 4.2.1 Product and Process Environmental Aspects and Impact
         Assessment

OP 4.3.7 Incident Preparedness and Response
         Operating Procedures (Including Preventive Measures)

Prepared by: John Engineer	Issue Number: 3
Approved by: Jane Manager	Issue Date: 10/25/95

4.0 Checking and Corrective Action

4.1 Monitoring and Measurement

4.1.1 Scope. The process described in this section conforms to ISO 14001, Section 4.4.1, Monitoring and Measurement. This process allows New Process, Inc. to monitor and measure its operations and activities related to significant impact on the environment and to track progress at meeting the goals established in its environmental policy and in the associated objectives and targets.

4.1.2 Responsibility. Each operating division is required to establish monitoring and measurement protocols for its operations to assess compliance with the company's environmental policy. The Divisional Environmental Protection function will provide help to its division in establishing monitoring and measurement systems, and will periodically assess their completeness and accuracy.

4.1.3 Process. Monitoring and measurement will include the means to evaluate periodically compliance with applicable environmental requirements and progress in reaching objectives and targets. Results will be regularly reviewed by management, at both divisional and corporate levels. Results will be transmitted annually to the Environmental Protection function for use in the company's annual environmental report.

4.1.4 Related Documentation. Operating procedures include the most recent version of:

OP 4.2.1 Product and Process Environmental Aspects and Impact Assessment

Prepared by: John Engineer	Issue Number: 2
Approved by: Jane Manager	Issue Date: 08/10/95

OP 4.4.1 Environmental Monitoring, Measurement, and Corrective Action

4.2 Nonconformance and Corrective and Preventive Action

4.2.1 Scope. The process described in this section conforms to ISO 14001, Section 4.4.2, Nonconformance and Corrective and Preventive Action. This process allows New Process to identify, investigate, and correct actual or potential instances of nonconformance to the environmental management system. Corrective or preventive actions taken to eliminate the causes of actual or potential nonconformance situations shall be appropriate to the magnitude of the problems and commensurate with the potential environmental impact.

4.2.2 Responsibility. Each operating division is required to develop and maintain procedures for identifying, investigating, and correcting actual or potential nonconformance situations. Each operating division shall implement these procedures for its operations, utilizing its monitoring and measurement system to assess compliance with applicable requirements and the environmental policy. The Corporate Environmental Protection function will provide guidance to the operating divisions in identifying, investigating, and correcting nonconformances and will periodically assess the adequacy of each division's nonconformance process.

4.2.3 Process. Each operating division shall establish a nonconformance process, in consultation with the Divisional Environmental Protection function. This process will include a means for identifying, investigating, and correcting nonconformances to the environmental management system. Results from this process will be reviewed regularly by management, including division and corporate management. Results will be reported to the Corporate and Divisional Environmental Protection functions as specified in Operating Procedure 4.4.1,

Prepared by: John Engineer	Issue Number: 2
Approved by: Jane Manager	Issue Date: 08/10/95

Environmental Monitoring, Measurement, and Corrective Action, including a record of changes in procedures resulting from corrective and preventive actions.

4.2.4   Related Documentation.   Operating procedures include the most recent version of:

OP 4.4.1 Environmental Monitoring, Measurement, and Corrective Action

4.3   Records

4.3.1   Scope.   The process described in this section conforms to ISO 14001, Section 4.4.3, Records. This process allows for the identification, maintenance, and disposition of environmental records, in conjunction with the document control procedures described in Section 3.5.

4.3.2   Responsibility.   The Corporate Environmental Protection function is responsible for defining this process and its associated system with input and support from appropriate groups, including Information Solutions, Administrative Support, and Legal. Operating divisions are required to use this process and associated system to properly identify, maintain, and dispose of environmental records.

4.3.3   Process.   New Process has chosen to use the information management system to store and retrieve current information electronically. Paper documentation originating from this electronic storage and retrieval system shall be considered potentially out of date, unless verified by the user as current at each time of use. Original paper documentation shall be kept in accordance with Operating Procedure 4.3.4.

4.3.4   Related Documentation.   Operating procedures include the most recent version of:

OP 4.3.4 Environmental System Information Management (includes
               Document and Records Control)
OP 4.4.1 Environmental Monitoring, Measurement, and Corrective Action

Prepared by: John Engineer	Issue Number: 2
Approved by: Jane Manager	Issue Date: 08/10/95

4.4 Environmental Management System Audit

4.4.1 Scope. The process described in this section conforms to ISO 14001, Section 4.4.4, Environmental Management System Audit. This process establishes a program for the conduct of periodic environmental management system audits. These shall:

- Determine whether the environmental management system conforms to Section 2.0 of this Environmental Management Manual

- Determine whether the environmental management system has been properly implemented and maintained

- Provide information on the audit results to management

4.4.2 Responsibility. The Divisional Environmental Protection function is responsible for periodically planning and organizing environmental management system audits for its division. Operating divisions shall provide staff and information needed to conduct such audits. Operating divisions are to maintain their operations in an audit-ready state at all times and may conduct self-assessments as they deem appropriate.

4.4.3 Process. The process utilized by the Divisional Environmental Protection function is detailed in Operating Procedure 4.4.4 and is available to all operating divisions. Audit frequency is based on several factors, including environmental impact potential, past audit and regulatory results, and degree of change in regulatory requirements. Audit results shall be reported to division and corporate management.

4.4.4 Related Documentation. Operating procedures include the most recent version of:

OP 4.4.4 Environmental Management System Audits
          Environmental Management Manual [this document]
          All Other Operating Procedures

Prepared by: John Engineer	Issue Number: 2
Approved by: Jane Manager	Issue Date: 08/10/95

5.0 Management Review

5.0.1 Scope. The process described in this section conforms to ISO 14001, Section 5.0, Management Review. This process provides for management reviews of operations and activities and for tracking of progress at meeting the environmental policy goals, objectives, and targets. A management review is to be performed and documented at intervals determined by management to ensure the environmental management system is:

- Suitable
- Adequate
- Effective

5.0.2 Responsibility. Top management will adopt procedures for the periodic conduct of management reviews. Each operating division shall conduct management reviews for its operations to assess conformance to the EMS. The Environmental Protection function is required to assist the operating divisions in establishing management review systems, and will periodically assess these systems as part of environmental management system audits. The Environmental Protection function is required to assemble companywide results for corporate management review.

5.0.3 Process. Each operating division shall establish a management review process, in consultation with the Divisional Environmental Protection function. This system will include an ongoing means for reviewing with division and corporate management the continuing suitability, adequacy, and effectiveness of the environmental management system. The management review process shall ensure that needed information is compiled to allow for a proper management review, such as:

- Corrective action and audit results
- Performance results for objectives and targets

Prepared by: John Engineer	Issue Number: 2
Approved by: Jane Manager	Issue Date: 08/10/95

- Changes in business environment that may influence policy, objectives, and targets
- New or changed legislation and other requirements
- New or changed stakeholder or interested-party expectations
- Changes in applicable technology, including work processes
- Organization's financial and competitive position
- Incidents, noncompliances, and nonconformances
- Monitoring and measurement data

5.0.4 Related Documentation. Operating procedures include the most recent version of:

Environmental Management Manual [this document]

All Operating Procedures

Prepared by: John Engineer	Issue Number: 2
Approved by: Jane Manager	Issue Date: 08/10/95

References

American Petroleum Institute, *STEP Brochure: We Must Demonstrate We Are Serious About Protecting the Environment*, Washington, D.C., 1993.

Chemical Manufacturers Association, "Guiding Principles," *Responsible Care®: A Public Commitment*, Washington, D.C., 1991.

International Chamber of Commerce, *The Business Charter for Sustainable Development: Principles for Environmental Management*, Paris, France, 1990.

National Response Team, "Criteria for Review of Hazardous Materials Emergency Plans," *National Response Team of the National Oil and Hazardous Substances Contingency Plan*, Washington, D.C., 1988.

United Nations, Conference on the Human Environment: Draft Declaration, Stockholm, 1972.

United Nations, Conference on Environment and Development, Rio de Janeiro, 1992.

United Nations, *Report of the United Nations Conference on the Human Environment: Stockholm, 5–16 June 1972*, New York, 1973.

Index

American National Standards Institute, 6, 16–17
American Society for Quality Control, 17
American Society for Testing and Materials, 17
Auditing (*see* Environmental auditing)

Bell, Christopher, 3
Bowers, Dorothy, 31
British Standard 7750, 10, 25–29

Canadian Standards Association, 16
Cascio, Joe, 107
Certification (*see* Registration)
Charm, Joel, 151
Checking and corrective action:
    EMS audit, 139–142
    monitoring and measurement, 130–137
    nonconformance and corrective and preventive action, 137–139
    records, 139
Chemical Manufacturers Association, environmental principles established by, 97–99
Conformity assessment:
    accreditation bodies, 85–87
    as it relates to ISO 14001, 89
    as it relates to ISO 14024, 91
    conformity assessment committee, role of, 91–92
    conformity assessment system, establishment of, 85–88
    elements of, 82–84
Connell, George, 16
Contractors and suppliers:
    management of, 126–127
    training, 120, 123–124

Department of Commerce (*see* U.S. Department of Commerce)
Department of Energy, 17

Eco Management and Audit Scheme Regulation, 25–30, 40, 66
Environmental aspects (*see* Planning)
Environmental aspects in product standards, Guide 64 (formerly ISO 14060), 53–56, 62
Environmental auditing:
    documents, 44–50
    ISO 14010, 45–47
    ISO 14011, 47–49
    ISO 14012, 49–50
    subcommittee members for, 44
Environmental labeling:
    documents, 56–59
    international programs, 56
    international trade, effect on, 59–60
    ISO 14020, 57–58, 62
    ISO 14021, 58, 62
    ISO 14024, 58–59, 62
    subcommittee members for, 57
Environmental management system standard (*see* ISO 14001)
Environmental performance evaluation:
    environmental management system, relation to, 50
    ISO 14031, 51–53, 62
    subcommittee members for, 51
Environmental policy:
    examples of policy statements, 104–105
    guiding principles for, 96–99
    key elements of, 100–103
    other elements of, 103–104
    top management commitment to, 99–100

219

About the Authors

JOSEPH CASCIO is Vice President of Environmental Management Systems and Managing Director of ISO 14000 Integrated Solutions for Global Environmental and Technology Foundations. He is the former Program Director of Environment, Health and Safety Standardization for IBM. Mr. Cascio has worked in the environmental field for over 14 years, and has a wealth of experience which includes national and international policy making, compliance management, and the development of standards for product safety and environmentally conscious products. Currently, he serves as the head of the U.S. ANSI delegation to the International Organization for Standardization (ISO) on International Environmental Management Standards, and is the chairman of the ANSI Technical Advisory Group (TAG) to ISO on environmental standardization. Internationally recognized as an environmental management expert, Mr. Cascio has played an integral role in the development of the ISO 14000 standards.

GAYLE WOODSIDE is the environmental engineering manager at the IBM facility in Austin, Texas. A registered professional engineer, she has over 15 years experience in the environmental management field and has expertise in hazardous waste, air emissions compliance, chemical management, wastewater-stormwater programs, and environmental technology. Ms. Woodside's professional affiliations include the Semiconductor Safety Association (SSA) and the Water Environment Federation (WEF). She currently serves on the Board of Directors for SSA, and is the co-chair of IBM's ISO 14000 Implementation Task Force. She is a former President of the Central Texas Chapter of WEF. Ms. Woodside has published numerous technical articles on environmental management and has authored several books.

PHILIP MITCHELL is Manager of Environmental Services with CH2M Hill in San Jose, California. He has helped define the environmental management field, developing a number of tools and approaches analogous to those used in the quality assurance area, specifically designed for effective environmental management and applicable to ISO 14000 registration. Mr. Mitchell is a registered professional engineer and a member of the American Society of Civil Engineers and the American Water Works Association. He is a former Environmental Manager for IBM.